BEE THE CHANGE

Bee The Change

IF WE PROTECT,
NATURE WILL PROVIDE

John Kotab

Contents

Introduction: How Long Will We Wait?

Modern agriculture is co-evolved with honeybees. If bees succeed, so do we. If bees falter, so too does that agriculture on which we all rely. Wild bees and pollinators, of equal importance, are also failing due to many of the same problems that face honeybees. We rely on them to carry pollen from one flower to another, which in turn fertilizes the flower, producing fruit. Everything from squash to strawberries, apples to avocados, requires this critical step of ecology. Some of our fruits and vegetables do not need pollination to produce our food, but if we are to get a harvest of seed for future food production, pollination is still critical.

As I spend my first official day of this trip in Ithaca, New York, I ponder on what the next three months will bring. After traveling by bicycle through New York, and then by train through Canada from Toronto to Vancouver, my pedal-powered travels will take me from the apples, cherries, and hazelnut orchards of Washington down to the almond, avocado, and orange groves of California. I will contact many farmers and other earth stewards in passing. Between these stops, hundreds of individuals each day will see an oddly inspiring man donning a bee-patterned helmet, complete with antennae, riding down U.S. Route 101 on a Vitamin "B" Road bicycle. The symbolism is unmistakable: I am representing the bees and insect life in general. I hope this does more than amuse those that pass. I feel I am doing more than "raising awareness" which can be hollow without spurring action.

This book will provide narratives and word-paintings to satisfy the adventurous soul, as well as interviews to educate and inform how we can heal bee populations and promote their habitats. As you take this journey with me, we will learn in our heads *and* in our hearts why bees and other pollinators are so important.

The best way to describe the role I've undertaken is *spokesman.* As with many of Earth's creatures, bees do not have a voice, or at least a language we can understand. As a follower of Jesus Christ, I find the imperative to cherish Creation inherent in scripture (see Genesis 1:28, also 9:1). Each species upon this planet is a gift from our Creator, each for specific purposes (see John 1:1-3). What a tragedy it would be to throw these gifts away. This trip therefore is one of the heart as well as the intellect. This trip is an attempt to give voice to pollinators, or to speak that which they cannot.

However, they *are* communicating with us. They are doing so strongly and relentlessly. They are communicating by dying, showing themselves legs-up by the millions, often ten-thousands at a time for the photo and video lens of legions of reporters. They are showing that their environment is toxic by wandering aimlessly and weakly, often never to return to their hives after partaking of nectar from a poisoned flower. And yet people still may not understand. The message is simple: All is not well on our planet. And in this case, where a large majority of the fruits, vegetables, and pasture forage we rely on for proper nutrition (whether direct consumption or indirectly via pastured animals) exists only due to pollination, the plea of the pollinator flies in the face of one of man's greatest self-deceptions: that we are separate and alien from nature.

The nature-care imperative in Christian scripture finds harmony with the wisdom-traditions of virtually all Native-American peoples. The indigenous wisdom found therein will be shared throughout this book, sometimes under the name permaculture – a way of interacting with the land for the purposes of food, fiber / craft, and medicine production that not only can be maintained for millennia, but is a relationship of care, of give-and-take, and not merely of extraction. We must also remember that while many native peoples have passed, many are still present and continue to serve Creation.

Being a living part of the Earth, we cannot harm any part of her without also harming ourselves. We must all see ourselves as part of this Earth, not as an enemy from the outside who tries to impose his will on it.
- John Fire Lame Deer, Lakota Elder

These poignant truths are most evident upon a walk in a State or National Park. Take, for instance, The Great Smoky Mountains National Park. You feel the cool breeze caress your carefree face. The sun shines down perfectly between trees on a calm day, a living kaleidoscope of green and light. Both are the free gifts of a forest structure, the world's most efficient climate regulator. A surprisingly harmonious feeling arises from the chirping cacophony of avian angels. The living soil is soft, cool, and black in your toes, while the animals, if you are reverent enough, can be heard rustling in the brush in all their variety, totally unaware of how beautiful they seem. All this comes from a healthy soil that has never been tilled or irrigated and is fertilized by each passing animal and insect. Divine perfection. A harmonious hum vibrates in the heart, and every breath gives not just oxygen, but joy.

Now, remove these things, one by one. The animals depart; the soil runs a sickly pale brown; the trees flee; the birds descend from their homes like the stars from the heavens. Hot days are punctuated by cold nights and the sun scorches. Is it any wonder that we value indoor square-footage (which seem to expand the more an ecosystem even as the one described above retreats) with such fervor? As the species diversity around us dwindles so do our places of sanctuary, where we may go and feel the presence of something larger, more powerful than ourselves.

At every turn of my writing experience, one core message continues to repeat: A field of flowers is just as much medicine to man as it is beneficial to bees.

It is important to realize that there is no need for condemnation of ourselves or others in this troublesome reality. We are born into a world and into a culture and received an upbringing that is independent of our choices. However, what we make of what we have been given is up to us. That is our opportunity. That *is our responsibility*. If we set our minds right and begin looking to nature as our provider, and not our antagonist, a beautiful healing could take place: both of our outer world, and the world inside of us.

1

Teach Me

"For managed bees, the honeybee colony loss rate was 54% in New York last year."

- Scott McArt, Cornell Research Scientist (2016)

I'd like to think of myself as a solitary bee, at least for now. Who knows what the west coast with its myriad of bicycle tourists buzzing down the U.S. Route 101 may bring? I found myself rapidly moving my legs back and forth in my sleeping bag this morning to generate warmth, just as eusocial bees do in their hives. I am still young in this journey, and I have so much to learn about bees and pollination. And I can think of no better place to start than in Ithaca, New York's Cornell University.

There were many opportunities to soak up the special beauty of this area during the day's travels. The environs of Cornell University campus contain marvelous waterfalls, chasms, and breathtaking hardwood forest which are sturdy on the steep slopes of the region and most glorious in all their autumnal colors. The ridges on the far side of Cayuga Lake almost glow in the late-day sun. I can think of no more beautiful locale for a place of learning, where nature is always the utmost teacher.

First attempt: the bottom floor of the Entomology Department at Cornell, directed there by a friendly gentleman who tells me that is where the "bee people" do their research. No luck. On the second floor of the Entomology Department, I have a riveting conversation with one woman who is studying Squash Bees (*Peponapis Pruinosa),* who finds my interest perplexing yet enjoyable. Then, off to the insect exhibit. I wasn't quite prepared for this: an entire library of invertebrates, three 20-foot rows of movable bookshelves twenty rows deep. Sadly, after asking to see the row containing the *Apis* genus, I was hurried out, being informed that to browse such a prized collection is only given to those who get permission and an escort.

Undeterred, I return to my first spot, hoping to find some pollen-filled enlightenment there. The "B" on the elevator gives me the notion that I am in the right spot. Again, the rooms are vacant. As I head back to the elevator, I see two men speedily passing and I call to them, not sure what to say, "Are you the bee people?" Professor McArt, the mentor to the student he is accompanying, says "Yes, we are!" I feel welcomed as he shows me a small Styrofoam cooler with a hundred tiny test tubes and says with an ecstatic hum, "We're about to do a project right now. Want to help?"

The experiment I stumbled upon was as follows: Trevor, undergrad at Cornell, is conducting research on honeybees, and has sampled pollen from fifty apiaries in the region. These will be analyzed through Liquid Chromatography (LC), and it was Trevor's job to prepare vessels for the LC. Scott McArt, head of the research lab, guided Trevor through the process while I listened. I soon discovered the profundity of this experiment: these samples are being tested for their pesticide content. Professor McArt explains, "The pollen in each hive contains about three percent corn pollen. Now, in regions like Indiana where there is nothing but corn for miles, bees will gather this pollen, and hence their hives will obviously have relatively high amounts of thiamethoxam, the common toxin with which corn-seed is pre-treated. However, in Upstate New York, where our agriculture is less intensive, we want to see if applications of these pesticides still pose a risk to bees."

Following his explanation, I want to better understand the nature of the experiment, and so I ask Scott for his motivation. He replies, "Research on pesticides in entomology is controversial because there is still little data on pesticides and bees." "So, there is still the image of being bee decline only happening in particular areas, and this experiment is to see if the problem is also in Upstate New York, and not a localized but a widespread phenomenon?" "Exactly. We need more information." I inquire, "Are there any studies being done on adjuvants (the "inert/non-active ingredients" that chemical companies use to increase target specificity and/or potency of their active ingredient)?" Scott informs me "Chris Mullin [Professor Emeritus of Entomology at PennState] has been finding that some adjuvants are toxic and can act as synergists [see example below], but we really don't know very much yet. There's not a lot of research done on adjuvants." I nod, and then reason, "It's just that the chemical companies have such a tight veil of secrecy over their inactive ingredients where those may be even more problematic." "Sort of. For example, it was found in a toxicology study that a certain Pyrethroid compound became up to a thousand times more toxic when a particular fungicide was introduced. Chris Mullin has been finding that some adjuvants are toxic and can act as synergists, but we really don't know very much yet."

This conversation turns back to bees specifically, and I ask, "What have you seen with the pollinator situation in Upstate New York?" Professor McArt informs, "We have 416 species of wild bees in New York. Several are just as abundant as honeybees. Of our 416 wild bee species, 53 are in decline (in other words, about 13% of New York's bee species are in decline). For managed bees, the honeybee colony loss rate was 54% in New York last year." Professor McArt also offers some good insights into the whole picture of bee decline, and what research is in the works:

"The main research motivation in my lab is to figure out why these declines are occurring. Most scientists agree that multiple stressors are responsible, including loss of habitat, pathogens, pesticides, climate change and inadequate management practices by beekeepers. We don't know which factor or factors are the most important, but we do have evidence that all are contributing to declines in pollinator health."

Talking with Professor McArt and Trevor has been most informative, and the hunger to learn drives me to my next locale. While the collegiate environment I left focuses primarily on the intellectual side of ecology, the next school provides additional focus on the emotional and spiritual: New Roots School in downtown Ithaca. At New Roots, high schoolers are given the opportunity to obtain a high school diploma while integrating interpersonal, intrapersonal, and earth care skills and awareness into their learning experience. The school has seen excellent results from this approach in striving to create not only scholars but whole humans.

I wasn't quite sure what to expect, but I got an email reply from Rebecca Cutter of Finger Lakes Permaculture and decided to show up, meet her and David (a fellow teacher), and see what I could find. I found a lot: a small charter school full of brilliant teachers with big hearts, and curious, open-minded students. The school is set up in an old hotel built in 1830, which gives the place a beautiful rustic feel. It reminds me of a time when children still knew the changes of the seasons, what each bird of the field sounded like, and the works of the community involved them, even if only in small ways.

Rebecca invites me into a classroom and encourages me to sit in on an object lesson. David leads a class of no more than ten high schoolers, having them write down things they show 'on the outside' and emotions they may not want others to see 'on the inside'. Those who did not want their papers read can opt out. David then introduces me, and the students loved what I was doing: trying to change some perspectives about our home, the Earth.

We go outside during what was one of the few breaks of sunshine that day and play ball, which is to say we crumpled up the papers we wrote on, threw them at one another, and randomly picked up another person's paper. After reading these anonymously, writing all the results on the white board, and highlighting the 'primary colors' of emotion (Fear, Happiness, and Sadness). We then discussed genuineness, and how we are all very alike in what we are feeling on the inside regardless of what we show on the outside. All told, it was a moving moment.

I am then given a short tour of the building and talk for a bit with Rebecca and David. We came to meet under the connecting influence of Scott Mann, producer of The Permaculture Podcast. I am told that the school satisfies state curriculum while teaching the students many important ecological lessons, systems thinking, and the rights of all living creatures. For example, during the required lessons on water, students identify metals and other elements in the periodic table that can end up in water, and learn how water gathers these materials, how pollution can occur through the water cycle, and what that means to us as human beings.

I also learn that a school project is currently underway to restore wetlands to the southern tip of Cayuga Lake. This tip has been embraced geographically by downtown Ithaca for centuries. New Roots is now embracing this lake with their hearts and giving back to it. The sweetest moment is when I ask Rebecca and David if they saw a lot of students succeeding at New Roots after struggling at the schools they previously attended. Their response was emotional and rang with purpose: "Yes, it's really something, and it's honestly the best part of being here."

I am here primarily to learn. So is Scott and Trevor. So is David and Rebecca. Only after we learn can we share with others. Only after we share something does that thing truly become ours. These people are here in Ithaca learning, sharing in the deepest ways, and being empowered so that one day, when they are tempted by economic or cultural forces to sell their birthright – A wild abundance, the environment their Creator has given them for providence – for a mess of pottage, they will have what it takes in their heads and their hearts to protect, so that nature can provide. My journey has just begun, and already it feels so rich.

2

New York

"We need to change our eating habits. We need to eat less of the government-subsidized staple foods [such as corn and wheat] and support the kinds of foods you see at farmer's markets."
- Denis, of Lakestone Farms (2016)

The journey became deliberately more difficult as I left Ithaca. This happened not by my doing but was precipitated by the universe as a necessary part of the refinement process. Ithaca was a grace-laden start. Geneva, New York brought bitter winds and rains. Palmyra, New York brought bodily disaster and sabotage of my most-important belongings. This was followed by a once-in-a-lifetime moment of sweetness in a grove of trees, enveloped in a spirituality that brought beauty and clarity of purpose to each shimmer of sunshine, each leaf, each step upon their soft decomposition. I then came to rest in the warm kindness of strangers that made the hard times seem to fade away into nothingness.

I made some good connections with pollinator researchers via the Geneva Research Farm and had a marvelous night at a small-town restaurant, followed by a good rest in my hammock strung between the steel posts of a railroad bridge. The next morning, I was on my way to Palmyra to visit the Sacred Grove, a holy place in the eyes of the Church of Jesus Christ of Latter-Day Saints, or that of the Mormon people. I was not more than two miles away before I was thrown into the air. My bicycle and its attached contents took plenty of damage as my shoulder pounded the grassy shoulder of Highway 21, now littered with the remnants of whatever had slammed my back. Other than whiplash, which grew agonizing as the hours passed, I came out of the situation relatively unscathed. I had been hit by the side-view mirror of a heavy-duty four-wheeler truck going fifty miles an hour. Both the driver and a witness stopped and awaited EMS which happened to drive by just minutes later. The witness, Karen, told me she thought I was a goner when she saw my legs fly above my head. A police officer handed me an insurance claim paper and Karen conveyed me and my belongings to the Sacred Grove.

She kindly awaited as I reverently paced the wood, looking into the autumnal spires of trees as if looking up at the stained glass of a cathedral. The leaves crumpled softly with every press of foot, each step seeming deeply meaningful. I found myself enveloped in a sacred space while receiving spiritual guidance I had long been seeking.

Karen then took me to the hotel where I would tend to my whiplash and wounds. Her kind assistance continued to bless me throughout the next day and assured that I was sent safe and sound on my way. That evening, I gave a call to a certain someone. We had been having long conversations for many weeks and were entertaining the idea of entering a serious relationship. Near-death experiences like I had bring one to boldness and lack of shame. During talking with her that night, I said, "I'm just so glad I get to be your boyfriend . . . If I can say that". She now looks back in laughter about how she was shocked but pleased at my rejoinder. Despite the pain and severe setbacks brought about by the accident, the second half of the day carries an eternal feeling in my memory.

Due to my injury, and subsequent need for a bicycle-repair shop, Karen continued to bless my life. In our travels, I happened upon the most excellent town of Canandaigua: a bustling place, in its small-town way, alive with parents and children enjoying a Halloween parade. Crosswalks gave the pedestrian right of way many times a minute, and the roads were decorated with facades upon business establishments and lovely murals. One in particular caught my eye: a sixty-foot-long alley painted to the brim with native wildflowers, such as Black-Eye Susan, Blue Lilac, and Sunflower. This alley led the way to the fruits of their presence: the town farmer's market. Because of these bee-friendly plants, which I saw in relative abundance in the region, these farmers have a wild-bee population to pollinate their crops for free.

I was able to talk with a few of the producers. Denis Lepel runs Lakestone Farm, a small-scale operation providing fantastic-tasting foods that are grown by healthy soil, not agri-business chemicals. His kids dance in place as he talks to customers, many of whom seem to be regulars. I am told that this is the last season for the summer market, but that there is a winter market, in which Lakestone also participates. I commend him for his full-diet production, where fruits, vegetables, and animal products are all coming from one farm. We get to talking about a renowned full-production operation in the region, Essex Farm, and *The Dirty Life*, a novel which chronicles its coming of age. I see in him a citizen who takes his role seriously and with reverence.

His life seems idyllic, however there is one challenge at his farm one might not expect: Denis is especially sensitive to bee stings. "I found this out when I was young. Just one sting on my hands will make them look like boxing gloves." You might balk, "Why work as an organic farmer then, having daily close contact with honeybees and wild bees?" I and each farmer at Canandaigua Farmers Market might reply: The mountains of squash, the pile of tomatoes, the bunches of turnips . . . these are the fruits of pollination. This man realized the indispensable role that pollinators play as man and nature participate in the co-evolutionary dance we call life. As awful as some of his experiences may have been, he understands that the ecosystem is our partner rather than our enemy, thus choosing to work closely with nature, our provider.

As Denis said, increasing our dietary intake of pollinator-dependent foods and decreasing our purchase of monoculture-sourced, machine-dependent crops is a rewarding and delicious step we can all take. This is a huge help for pollinators because monocultures such as corn, wheat, and soy are a direct affront to biodiversity, and local produce, especially that found in farmer's markets, reflects a model where a bounty of food is produced through the virtue of a healthy ecosystem (replete with both cultivated and wild flowering species), not agricultural chemicals. Bees and other pollinators need a wide variety of food species that bloom throughout the year for their pollen, not one or two bursts of pollen from a single species each year. In areas where monocultures dominate, year-round pollinators are seldom found.

Sunday brings more opportunities than I can relate. In the morning, I attend church at a local Latter-Day Saint congregation. A note about my previous religion: While I grew up LDS, or as some would say, Mormon, I have gravitated away from it after 2020, favoring to identify with Christianity as a whole. As I said, while I am chiefly a follower or Jesus, I still respect all of the wonderful wisdom traditions His teachings have inspired. While I no longer identify with the Mormon wisdom tradition, I also honor where I come from and recognize its positive effects on my life.

The first man I shake hands with, Paul DeGraff, happens to be a hobby beekeeper. Next, I meet a sharp young man who, according to his mother, was an entomologist from the age of two. Then enter Pat and Gary West, the sweetest couple of more than forty years who do part-time work at the Bishop's Storehouse, a church-run food pantry that collects, cans, and distributes food to the needy. These two do it out of the goodness of their unified heart as a volunteer service. I learn that Gary was once an apple farmer. He, with passion that suggests he never retired at all, describes how much pollination is desired.

"You see one bud come out, and that becomes the king blossom, and later four blossoms emerge, surrounding the king blossom. It's good to have one or two of those blossoms develop an apple," Gary animatedly tells me. I add, "Any more than that, you end up with a lot of sugarless orbs rather than the tender sweet apples that people want." "Correct," Gary replies.

I think that short hour of social time at church had me talking with at least two dozen people about my cause and website. Afterwards, Gary invites me to spend the night at their place. During the ride there, Gary and I get to know each other better. As he walks into the house to greet his wife, he declares, "You'll like him. He is one of us. He eats like us." I felt warmly received in that moment, and again when they prepared a hearty helping of tomato soup which they had canned from the garden harvest. We then broke bread together and had a wonderful evening visiting.

This was honestly the best part of my time in New York. So much sincerity, sharing, and the indescribable sensation of the true self emerging. The West's even introduce me to their family during an evening potluck, where I am further able to share my love for nature. Pat brought a pie made with blueberries and currants from the garden. That pie, not to mention all the rest of her cooking, was remarkable. Pat and Gary showed me true abundance. Their love for one another bloomed as only long-time romance can. Every prayer over food ends with an "Amen" and a kiss between them. Their patio bursts with the crimson, turquoise, and pale-yellow hues of Winter Squash. Their root cellar sports six-foot high shelves stocked in mason jars, with more than fifty jars of homegrown tomato soup. "Now *this,*" I proclaim, "Is social security." I'm glad to say that pollinators are abundant as well in their region of Albion, although their pollination needs are few and each region has differing degrees of bee decline.

After leaving the West's home, I experience the most noteworthy manifestation of my "inner bee". A most remarkable field of sunflowers, under-cropped with late daikon radish, comes into view. A pollinator paradise. I have long been interested the agricultural practice of cover cropping and this was a fine example of it. Upon large sunflowers such as these, dozens of pollinators (and other bugs. The amount of nectar and pollen available is immense) can be seen at once. I take out my tripod for a photoshoot.

One sunflower catches my eye and, fascinated, my hand draws near to admire the texture of its ripening seeds and silken ray flowers: an alluring and sensory-rich encounter. This interaction seemed a perfect human representation of a bee's-eye view of the world. The picture (entitled *Sunflower Rendezvous*) that resulted is shown on the back cover of this book

On my way to Canada, I see two sides of New York agriculture. I pass fields of lovely apple orchards, their fruits blushing with the late fall, awaiting the frosts necessary to obtain their greatest sweetness, where they will then be carefully gathered by human hands. In contrast, the Erie Canal trail borders vast tracts of corn, also awaiting harvest. I watch two big red combines move through one field as 18-wheelers await their bounty. As these machines churn and tear through literally tons of corn grain, I stand for a half-hour in awe as a silent witness to this 21st century industrial act of sheer power. As I ponder on the spiritual energy of each method – mechanized, heavy-handed corn culture and bee-friendly, human-reliant apple culture – it becomes all the clearer which future I want to support with my dollars.

Remarkable gifts from the universe mount up as I approach the terminus of America. Lockport, where the Erie Canal bicycle trail and I part ways, shows me once again incredible ingenuity and horsepower, this time of 19th century industry. I climb more than a hundred feet over the course of multiple locks that convey barges and boats "uphill". The remaining ride offers countless opportunities, with porch lights and festive decorations hung, to stop and fill up on a ration of Halloween candy, human smiles, and kindness. The spirits are out tonight. A keen magic rides the air and conveys me to the falls. And not just any falls, but *the* falls: Niagara.

Enormous mists and sounds fill the air as water crashes into rock and yet more water, doing the slow, grinding work of eroding away rock and producing sediment, as it has for millennia. My twenty-seven-year-old eyes were just as captivated as when I came here at the age of ten. I think on what a woman told me back then, as we all looked on: The works of man eventually tire the eye upon repeated looking, but nature -- the work of God -- never tires the eye, or the soul. Even at night the falls are larger than life, with careening foam lit up in the ambiance of city lights in some places and by multi-color floodlights in others. I steal away into the shrubby forest to set up my sleeping bag. I am comforted by the swift waters rushing by but am spurred back out amongst the dissipating small crowds. It is a brisk evening which gets to me the more my body cools and my mind tires. As I make my last venture to a lookout point, three monks enrobed in orange appear and I am drawn to them. We look out at the grandeur together and converse. They are from Tibet and are visiting this last week of October.

They listen as I relate my sojourn for the bees, for all pollinators, and for all insect life. These spiritual guardians then give me my last gift of the day: a pronouncement that my work will be guided by Spirit and yield much fruit. It was an impactful endowment that to this day gives me chills – The paths of the lone entomology pilgrim and three holy men aligned by destiny. The Spirits are indeed about, churning and flowing us towards our highest purpose. I now know mine for this season of life: to lend hand, voice, and heart in protecting Creation.

3

Transcontinental

After I went through New York – and after New York went through me – I pass the misted portals of Niagara Falls, cross Rainbow Bridge and hop on busses and trains bound for Toronto: a flurry of car-lights, cyclists, and railcars. Finally, there arrived an experience that I have wanted to have for years. Via Rail Canada: this would shuttle me across the prairies, snowy peaks, glaciers, and vast expanses of Canada. I met a remarkable palette of humans in this trip. Trains are beautiful places to meet others. I highly recommend this trip provided you get fare for the sleeper.

The following interview took place on the Via Rail Canada Line, heading west from Toronto. We were somewhere between Saskatoon and Edmonton. A warm spell had passed over this region of the north, and so my jaunts outside of the train were uncharacteristically cool during the evening, but full of warmth and sunshine in the day. This was a fitting parallel to the open and enriching interaction that took place.

As some of my fellow passengers ask about threats to bees, I express my concern over monocultures, which often necessitate high amounts of agricultural chemicals. Glyphosate is mentioned and I relate that though it is assumed to be benign, it disorients bees and harms them significantly. One passenger's ears perk up at the mention of the Round-Up active ingredient and chimes in, "I'm in a bit of a unique situation: I do work that uses Glyphosate to *prevent* monocultures." Intrigued, I inquire further, and he replies, "To prevent invasive species. They are a big problem in Washington State".

The bright young man who introduces this welcome complexity to the conversation is Alex Sturbaum. Alex works with AmeriCorps in cooperation with the state of Washington. He is a restoration worker. Washington State is very interested in his line of work because it helps to increase biodiversity, repair habitats that are critical for healthy salmon populations, and combat environmental degradation, including erosion and – central to my interests – pollinator decline. Through this interview, we conclude that monocultures precipitate immense losses in the environment whether directly through agriculture or indirectly through invasive species takeover.

What follows is an in-depth look into an earth-lover's labor amongst the riparian areas of Washington. We launch right into the discussion, beginning with farm buffers.

I am fascinated by the opportunity to see things in a whole new light regarding Round-Up and ask him for a brief introduction and how he got into conservation work. Alex relates, "I've always been passionate about environmental restoration and leaving the world in a better condition than we found it. I moved out [to Washington] following my then-fiancée who was doing her PhD at the University of Washington. I heard about the Washington Conservation Corps (WWC) and saw an entry-level position in conservation. I joined it and have been working with them since. As for what we do, and as it relates to increasing biodiversity: We work along both public and private property. My crew works on private property – both farms and people's backyards. We are a voluntary agency, so we're all carrot and no stick, as it were.

Our work is two-fold: We remove invasive species like Japanese Knotweed and English Ivy. These are species that come in, choke out native vegetation, and establish a monoculture. We then plant native species along riparian areas which usually have salmon. That's part of why the WCC is fairly well-funded, because salmon are pretty significant [for Washington's fishing industry]. So, salmon need streams that have lots of shade to keep temperatures down. They need large woody debris their juveniles can hide in, so we establish buffers of native vegetation along these streams, and we'll monitor these areas over the next couple of years and continue to eliminate any invasive species until the native vegetation can hold their own. In addition to natural streams, we'll frequently work on agricultural ditches on sides of farms. The farmers will cut the agricultural ditch straight and narrow."

I ask if this to keep the agricultural land relatively dry and what their size is. He replies in the affirmative and tells me they are quite large. "Those are often stream channels that have been diverted and frequently salmon use these agricultural ditches, so we try to clean them up and make them a more suitable salmon habitat. The law requires a twenty to fifty-foot buffer of native vegetation around the agricultural ditch. The WCC crew I work on currently works for Keen conservation district, a non-regulatory agency that works cooperatively with farmers. And, of course, any buffer we put in takes away from the farmer's land, so it's give-and-take."

> Washingtonian farmers feel a kinship to the earth and often realize there is more to consider than the bottom line.

Alex continues, "We're fortunate that in Washington State a lot of farmers are actually very passionate about preserving nature." "They feel that sense of earth-stewardship, then. That's beautiful. Your talk about the buffers reminds me of what I read in David Holmgren's *Permaculture Pathways and Principles Towards Sustainability*. One of the concepts is to value the edge and to recognize that incredible things happen in the margins, both biologically and socially. He said that in Old England there were mainly small farms, and that the edges between farms – the hedgerows – were important biological structures. As the farms got bigger one of the big reasons for environmental decline is that there was less edge. It's good to hear that a lot of Washington farmers are on the ball relative to care for the environment. They realize that they need to invest in the future. They want the land to be arable in twenty or thirty years." Alex concurs and adds, "I think that sort of long-term thinking is [being adopted at a] slow [pace] but it is beginning to catch on around the nation and around the world as the scientific data mount up."

I then ask, "What kind of work does AmeriCorps do throughout the country?" Alex animatedly tells of the organization in which he serves. "AmeriCorps is an organization that does all kinds of work that needs doing. They work in schools, in low-income neighborhoods, etc. They build houses, do disaster relief, etc. The WCC is one of many throughout the country that work under the general umbrella of AmeriCorps but partner with state agencies. The WCC is a partnership between the Washington Department of Ecology (WDOE) and AmeriCorps, and gets their funding partly from AmeriCorps, partly from WDOE, and through hiring out to state agencies. I've been on restoration teams for two years now."

Following his explanation, I ask whether he has seen his work increasing diversity and benefitting pollinators. He explains that because the crews he works on does the restoration work and then moves onto a new site that needs the same rehabilitation, he has not. He explains, "And the restoration of the site occurs on a timespan of five to ten years." He then opines, "I would say that our work [along the farm buffers] creates that edge-benefit you were talking about, and these are microcosms of a diversified ecosystem that I think pollinators would thrive in."

Even in conservation work, worthy desires face the complexity of our world.

Alex regretfully adds, "It *is* disheartening to see a bunch of honeybees buzzing around Japanese Knotweed flowers, especially when you're there to douse the plants in Glyphosate in order to prevent erosion on the streambanks." Doing my best to maintain objectivity and allow him the space to tell his story, I maintain curiosity and inquire further. Alex continues, "See, knotweed grows very tall but has a brittle root system, and it dies off in the winter and flows downstream. These broken-off parts can re-root further downstream. It will basically choke out whatever native species have deep root systems on the streambank. When it dies back in the winter, there's nothing holding the bank, and so larger amounts of silt are dumped into the stream." Enlightened, I support his position. I understand that all species of salmon are keystone species in the coastal and interior regions of the Pacific Northwest. "That's a huge problem. I know that salmon have requirements for the turbidity of the water as well as temperature." Alex affirms this, "Right. They need clear, cold water."

Along this vein, I ask further, "What other options have they tried, but have found Glyphosate to be better?" "We've tried several different herbicides on Japanese Knotweed. We use Imazopyr because it goes into the roots, lasts longer, and is more effective at taking out the plant on a yearly basis. In a highly infested space, we might mix Glyphosate with Imazopyr. Triclopyr is also used."

As I assimilate the strange fact that weed killers are being used to re-store biodiversity rather than destroy it, I muse, "It's so funny. We live in a complex world. There are so many different cost-benefit factors to weigh. Luckily, by the time these areas are in bloom and the bees move in, this chemical has decayed and is no longer active." Alex responds, "Yeah, and we do try to take pollinators into consideration when spray-ing. For example, we don't blanket-apply. We go out with backpacks and target-spray as needed. We also time our spraying when there are few bees out. We do what we can, which is regrettably sometimes not a lot." I again offer, "We live in a complex world. It's really fascinating what you said that you are spraying to prevent monocultures instead of spraying to prop up monocultures."

I change gears. "So now: You've done restoration on a large scale. Once the weeds are gone, what is next?" "We manually plant. We sometimes do bare rooting, and we sometimes do live-stakes. So, with stuff like willows and cottonwoods, they propagate vegetatively. The WCC has the benefit of having people working for them between the ages of 18 to 25, and we're also cheaper than goats!" We each have a good laugh, and he continues, "We ran the numbers once. We found that it's actually cheaper to have a WCC crew to come and remove your blackberries than to bring out someone with a herd of goats." "That's fantastic." "We take care to plant them properly by hand." I ask, "And the planting: When does that happen? The moist season?" He affirms that winter is the planting season and that as I travel, I may see a crew or two and to look for those with yellow hoodies and blue hard-hats.

This is the moment you all have been waiting for. Once we learn and are drawn closer to the natural world, we feel inspired, but that feeling can diminish if we don't take action. Alex gives some great advice.

The conversation has been enriching and eye-opening. I ask what we do as the general public to get on board with restoration and play our part. He answers beautifully, "Learn how to recognize and how to deal with invasive species. Most counties these days have posted online a list of noxious weeds and best practices. Consider planting native species in your yards and gardens as opposed to what just might be pretty. It's always good to give local pollinators native plants. And before you do anything on your property, investigate the effects it will have. For example, pick up your pet waste and don't wash your car in your driveway. There's a lot of things you can do to make your environment a little healthier." I add, "The greywater movement is picking up a lot of speed. It is switching a lot of people to the biodegradable detergents where they can re-use their water, including for their lawns." "Absolutely, and I think citizens are getting a lot more aware."

Alex continues, "I also think we need to put a little more pressure on industry. Frequently, while the waste/damage that's done by citizens in their personal lives is a substantial number if you add it all up, industry at least equals it, and often outstrips it. It's two-fold: analyzing what you do at home and putting pressure on those who provide us with our products and services." I then relate an insightful assessment of the business world told me by a couple I met at church years ago. "You can measure whether you are a successful businessman by improving society and making people's lives better, or you can measure it by how your bank account is swelling. You can get the money by accident, but it's much harder to get the improvement to society by accident." Alex concludes, "And I guess the trick as consumers is to make sure that the one equals the other."

You can learn more about the Washington Conservation Corps at ecology.wa.gov. I think it is important to re-emphasize that each of us has a part to play, and even small zones of native flowering plants -- including the willows and cottonwoods that Alex's team placed along stream banks -- provide buffers for pollinators. This work can also provide additional benefits, like creating river habitat more suitable to salmon. These beings are critical to not only the region's economy but also culture – that true culture that always arises from an intimate connection with the Earth. This feeds us spiritually and brings us together in meaningful ways.

I would add that herbicides are not the only option for performing a "reset" on land you wish to restore. Controlled grazing, flame weeding (when appropriate), smothering with a tarp or heavy-duty opaque plastic (more effective in the warmer, sunnier months), or a combination of these in succession produce wonderful results. A buffer for pollinators can be as small as a riparian area restoration project, or even as minute as your backyard. Finally, not all non-natives are invasive. Many trees, shrubs, and annuals which cohabitate beautifully with their neighbors have been in the region for as little as 80 years. Some ecologists refer to these plants as species with have naturalized. They need no removal efforts.

Another individual I met while upon the train deserves mention. He was around ten years older than I. Though his name still escapes me, he left an indelible impression. A bicycle-culture entrepreneur and natural salesman, he and I had many rich conversations, but one stands out. He always came off cool and collected, but one thing he said carried an extra air of certainty and calmness. It was this:

"Ah, your birthday has just passed, and so you are twenty-seven and there is a special combination of celestial powers rising in your life. This alignment comes only once every seven years. There are things the universe will try to teach you in this season of your life, which if you ignore, seven years will pass before you can again assimilate into your life that light and understanding." This was in response to our conversing about how the Buddha and John Muir had massive paradigm shifts in their late twenties where rare life events strongly impelled them to live out their highest purpose.

Perhaps now is that same time in our collective consciousness as mankind. Is it possible that it is not the end of the world nor irreparable ecological collapse that faces us, but rather a golden opportunity for spiritual ascension that is passing us by as like a comet, which may not return until generations later?

I asked in the beginning of this book, "How long will we wait?"

I ask it again.

4

Hives and Hooks

"I think the bee plants are very useful. Every region has [native] plants that are bee-friendly. We can put aside a part of our garden to plant these. Many of them don't need any attention and come back year after year."

- Mark Urnes, of the Washington State Beekeepers Association (2016)

A week ago, I looked from the southern edge of Victoria to Port Angeles. Now, from the ridge above, I see the lovely crabbing town of Port Angeles, Ediz Hook stretching its protective arm across the bay further in the distance, and Vancouver Island nearing the horizon. The uniqueness of this part of the world is unmistakable. Sporting multiple peaks nearing 8,000 feet above the sea yet only 10 to 20 miles from the shore, the Olympic Peninsula contains some of the wettest and driest areas in the Pacific Northwest. Sequim (slightly east of Port Angeles) receives an arid 22 inches a year, and Quinault a whopping 122 inches. This is due to what is called a rain-shadow effect, and this 'shadow' carries all the way into the eastern tip of Vancouver Island (Canada), affording the same lovely climate that Port Angeles enjoys. Only the Hawaiian Islands and Death Valley witness this effect as dramatically in the United States.

However, the Olympic Peninsula is in a very temperate climate, and contains little industry and almost no commercial agriculture. Glacier-fed rivers burst with salmon in the almost-Mediterranean weather. The adjoining Puget Sound contains myriad islands, each with their own personality, climate, and culture. This is the peculiar and lovely environs in which I basked for three days during the time I met and interviewed Mark.

Mark Urnes, of Port Angeles, Washington, is the Education Trustee of WASBA North Olympic Peninsula Beekeepers chapter and has been keeping bees for decades now. His property sits at a delightful location in the foothills of the Olympic Mountains, with the eastern Olympics (including Mt. Angeles and Hurricane Ridge) as a backdrop. His small grove of apple trees just outside of his fence overlooks the ocean below. On one of the sunnier days in November, he shows me his twelve hives. I get to see his pollinator garden, complete with Borage, Mint, Bee Balm, and others. His honey collection centrifuge, beekeeper outfit, beeswax candles, and jars of golden amber honey are all lovely to see.

After this introduction we sit down, and I get to learn what this man is about. I begin by asking about the history of beekeeping in this part of Washington. Mark fills me in. "Well, there is not a whole lot of ancient history. Beekeeping in general didn't come to the west until the mid-1800s. It really came during the gold rush in 1850. The bees themselves, the Western Honeybee (*Apis Mellifera),* were not native to America. They were imported in the 1500s and spread across the continent up to the continental divide, where they couldn't get over because of the elevation. There were no honeybees at all on the west coast until they were shipped around Cape Horn."

Mark continues, "Honeybees have not been in our neck of the woods until the turn of the 20th century. They don't do as well here because of the cold maritime climate. They just haven't adapted to the climate. There is actually a movement afoot to do more local raising of queens, so that they are more adapted to their own area. Currently, we try and use bees that come from a similar geographical location. Ours are Carniolan bees, they come from mountainous regions of central Europe. They come from a climate typical to what we have here, the thought being that if they do well there they will do well here."

I am impressed by his knowledge and am curious about his background. "So, tell me a little bit about yourself, your personal history. How did you get interested in honeybees?" "Seattle has classes in urban horticulture. We were right in the city of Seattle, so it wasn't an ideal spot for keeping bees. We waited until '94 when we moved out here. In '95 we had our first hives, and we've had them ever since. We've never had more than a dozen hives. That's a good number for me." I added, "Any more than that, and they start to own you. Its work." "It becomes a business, and losses are bigger. Losing a third of your hive when you have twelve is not a big deal. When you have a 100 hives, and you lose 35: That's a lot, at 80-90 dollars a hive. That becomes a big financial deal, and that's just for the bees themselves. That's not including the woodwork. It runs around 135-200 dollars for the wooden parts for the hive." Remembering the Warré pattern Mark showed me, I ask if it is a cheaper option for hives. "Yes, it is. One of the advantages of them is that you can go get a bunch of 2x12s at the lumber store, cut them out and put them together yourself. That's a big plus. It's a bit difficult to extract the honey, though. You actually have to crush the combs and drain the honey out that way."

I presume this means more work for the bees, and Mark affirms. "They have to make wax every year, and a lot of their energy stores are taken up with that process. That's one of the reasons you want to preserve as much wax as you can. I only take the capping from the wax, and I leave the rest in the hive. The wax has about a five-to-seven-year lifespan, after that it becomes so dark that it's unusable. They're finding that the wax absorbs a lot of the chemicals that the bees come into contact with, so it's generally thought that hives ought to be changed out every five years."

As our conversation drifts towards honeybees' challenges with environmental problems, we talk a little about the special nature of this corridor of beekeeping, and Mark shares a deep insight.

I then inquire of Mark regarding University of California at Davis. His response greatly expands my interest, and he connects it with our previous discussion about man-made chemicals in the hive. "It probably is the center of bee research on the west coast because it is centrally located and in the heart of almond country. And as you know, the relation of bees to almond farming is huge. I think they study all aspects of honeybees, including longevity issues with the queen, effective pesticides which don't harm the bees. It all becomes really important there. Here its theoretical things because we don't have pesticides here much."

I excitedly expound, "Right. It's neat that I came across this environment because it's somewhat of a control group. If I want to get an idea of, 'just how damaging are these pesticides in a given area,' and you're already losing a third of your hives even without pesticide problems, then that's very telling, I think." "Which to me says that maybe pesticides aren't the main problem in this region." "It's always a confluence of factors. But people don't want to hear that bee decline is a confluence of these dozen factors in these specific regions. That doesn't make headlines, right?" Mark smiles, "Correct. And my personal feeling is that it's something fundamental like the genetics of the bees. The gene pool has been reduced."

As we turn to talking about Mark's role as an apiary educator, the prior information gathers meaning as the dew from heaven upon his garden flowers.

I continue, changing the pace, "What I'm really interested in is to hear about your work with the elementary schools. What are some things you do?" Mark gladly explains, "The first step is to get people comfortable with bees. A large number of the people who approach the booths at the fair will just recoil away from the bees. 'Oh, I hate bees. Bees will sting me.' They get this partly because people get stung by yellow jackets. Some people just have this unfounded fear of honeybees, and they pass it on to their children. So, if you can counteract that, so people realize that bees aren't out there to get us, that they are just trying to get by, I think that helps a whole lot."

I inquire, "On a given school year, how much outreach do you get to have?" "I will probably go to a dozen classrooms in the course of a school year. We've done some already. In the springtime, when the time comes to talking about insects, we'll have another round of outreach. Our main interaction with the public comes from the fair, and the kids love it. We've had a presence there for about twenty years. We bring the observation hive which is basically a glass-walled hive so they can see the bees, the babies, and the workers going about their business within. The kids can listen to the bees too. They put their ears up to the glass and they can hear them. It really makes a difference. You can just see their faces light up. We've also been taking the drone bees, because they don't have stingers. You can hold them on your hand without risk of being stung. We'll tie a tether to the drones and let them walk around on peoples' hands."

The idea of people having their fear of nature ameliorated fills me with hope and delight. "I think that's all it takes. People just need to let nature touch their hearts, and they realize it isn't something to fear. How have you gone even further to help children realize that bees are our providers? What have you been able to do for that?" Mark quips then explains, "I think by just pointing out that a third of the food we eat is profoundly impacted by honeybees in some way. You'll hear people claim that without bees that life for humans will end, but really the food is going to become a lot more expensive. And at the same time, we have some four-thousand other native pollinators around nationally. They just don't have the workforce to pollinate to the degree that we need."

Referring to his education outreach, I ask, "Do you get contacts from people via emails or letters showing appreciation?" Mark delightedly replies, "We get a lot of booklets from the classes we visit. I love getting those from the students." Mark gets up and enthusiastically grabs a booklet from one elementary school and shows it to me. We turn its pages . . . Mark says that teachers tell kids to tell one fact that they loved along with the thank you, and to draw a picture to go along with it. I read some of the entries. "We appreciate you Mr. Urnes and your bees. I did not know that queen bees sting nothing else except other queen bees." Mark smiles, "The facts they grab onto is so interesting." I continue, "I did not know they make red, green, and blue honey too." "Yes, you know where bees will find a factory and collect the dyes along with the sugar syrup," Mark explains.

I am touched and say, "I bet this makes you feel so good reading these. You have facilitated children being exposed to nature at a young age. They learn how nature is our provider, and not something we have to compete against. There's some truth that we are having to compete with nature, but a large portion of that comes from us trying to ameliorate the effects of our own foolish actions. We're throwing nature out of balance and trying to restore it back to normal and providing these marvelous solutions, where the solution may have been to not throw it out of balance in the first place." Mark echoes my sentiment. "We think we're doing well at the time, but hindsight will tell us otherwise much of the time."

In every interview I conduct, I strive to immerse both myself and the reader in the world of pollinators, to speak that which they hope we would know. In a differing way each time, I ask the interviewee to leave a small parcel of experience-won wisdom on how we can help our pollinators' situation. I ask, "What's something we can do to ensure a more bee-friendly future?" Mark says, "I think the bee plants are very useful. Every region has plants that are bee friendly. We can put aside a part of our garden to plant these. Many of them don't need any attention and come back year after year. Even on a large scale, there's lots of planting of bee corridors, large areas near monoculture regions." I remember something relevant from a previous interview and add, "In Washington I believe there's a law that requires twenty to fifty feet of buffer of native vegetation between a farm and a tributary or river."

I wrap up by asking, "And with education, you've had the opportunity to teach in a very hands-on way, but what's something I can do, and we can do, to promote that positive view of nature and bees in particular?" Mark ponders and responds, "I think the more you learn about bees, the more interesting they become. I've been learning for about twenty years, and still, I'm learning more about them all the time. You never reach the endpoint. It's like they're full of mystery."

A world of mystery. I myself have felt this magical draw as I've entered their realm. As Mark and I discussed ecological buffers, the Christian principle of tithing came to mind, how many give ten percent of their income to assist in the ministering of the poor and disadvantaged in this world (See Deuteronomy 26:12). Certainly, the same principle could find fulfillment in our land use. Could we dedicate ten percent of our land to providing food and habitat for the creatures of God's world – which he called good (See Genesis 1:12, 21, 25, & 31)?

Could we return a portion of all he has given so that all natural beings can thrive?

And regarding my commentary on humanity's 'hard work' at competing with nature, Mark later informed me that Ediz Hook, a famous sandbar feature of the crabbing town, is reclaiming its importance in the eyes of the county. In the previous century, many of the rivers feeding into the bay have been dammed, thus reducing sand deposition onto the sandbar and threatening crab habitat. In recent years, the county has been welcoming nature back into the lives of the townspeople. Many of the dams are being removed to allow this sediment to feed the Hook, so that the crab population regrowth can feed the economy and the people. It is supposed that this should also help revitalize salmon populations in these rivers.

This will, in time, invigorate the forest as salmon begin to again give their gift to the Olympic interior: their bodies. This mating ritual of returning to the headwaters of their birth renews the forest with marine minerals and other nutrients as they become part of that interior's food chain. This removal project comes at a price-time tag of 300 million dollars and three years. Perhaps this is the style of "hard work" we can learn to avoid as we return to a natural abundance. Perhaps we too can learn, with our hearts, to consider pollinators, salmon, and all of nature sacred.

There is another aspect of the indigenous peoples of coastal Washington that stand out to me. Just as some Christian traditions speak of the importance of covenants (a spiritual contract between God and man), the Lummi, Tulalip, and more than a dozen other tribes here speak of their covenant with Salmon Woman: that if they respect and uphold the rights of the salmon, the salmon will always feed the people, as well as nourish the rest of creation. As I say in the book's subtitle, so it is: *If we protect, Nature will provide.* May we begin to remember these ancient ways and re-invigorate our culture with their lessons.

5

Snapshots: Vancouver to Portland

Saturday, November 12, 2016

After days under the pall of cloud and mist in Vancouver, my poorly chosen camp intersecting the haunts of the homeless and drug addicted, I choose to depart for a time. Though the repairs made on my panniers in New York were helpful, they provided a short-term solution which utterly failed as soon as I landed in Vancouver, making my mood to match the melancholy cloud cover. I emerge out of its disheartening shadow into a burst of light, beautified by the dripping remains of the cloud behind. Just an hour before, I gazed at cars, bicyclists, and even semi-trucks flying out of the hull of a ferry boat. It was an other-worldly sight to my fresh eyes, and I was most stimulated by the bicyclists pouring out after descending the ferry's steep ramp. Just fifteen minutes after boarding, all of us aboard witnessed a stark change from cloud to sunshine.

Our ferry is passing between the first pair of islands, near Matthews Point, on our way to Schwartz Bay of Vancouver Island. I have left one shadow and entered another: the rain shadow of the mighty Olympic

Mountains. Soon after the ferry lands and it is my turn to race out onto shore, the electrifying comradery of adjacent cyclists quickly fades as we all pedal off to our separate ways. As that feeling fades, so does the day, and the sweet-smelling mists of dusk accompany me to my shelter that night.

Monday, November 14, 2016

At Sooke River Gorge, I am among giants. Sitka Spruce surround me as I float by. Manzanita, much of its peripheral bark fallen, holds the appearance of burnished bronze rather than of wood. Exhalations of vast forests creep along rugged valleys. The only thing greater than viewing all this from old railroad grade paths is traversing them via bicycle. The Galloping Goose access trail, as it takes me along Happy Valley Rd and the coast of the Strait of Juan De Fuca, offers such an experience. I plunge thru sections carved into lichen-laden rocks, careen over railroad bridges perfectly weathered with moss and the years and feel the rapids hurtling downwards and outwards as my blood rushes inwards.

The trail is exhilarating, but treacherous. The multitude of autumn leaves creates more than enough detritus to slide me off my bike seat at one point, placing wide scrapes over my left arm and leg and scattering my belongings in every direction. I cross paths with an old transplant from England, happy and healthy with his bicycle by his side. I asked if it was expensive to create the 55km trail, with having to remove all the material. He explained that in the early 1900s, railroad was laid to access the mighty trees of this region. Even a gold-mine opened shortly. By 1960, shipments ceased, and after the railroad days it didn't take long for all the track to disappear! People took it and crafted its quality steel for their own use. As we parted ways, he places a band-aid on my morale, jovially pointing to my fresh road rash, saying, "Oh, you fall down and scrape yourself up, you get up and you keep going, that's how it goes."

After a night at Potholes Provincial Park, I receive word that the clips for my bicycle panniers have arrived. I am relieved to know that the results of the New York accident are behind me, and I can continue my journey.

Tuesday, November 15, 2016

I've broken my old record for miles in one day using commute-transit -- 132 miles, or 213 kilometers: an all-day venture. Fixed bicycle bags mean freedom from Vancouver. I applied this freedom without reservation, belting out a hearty chorus or two of "O Canada" as I neared Sumas, Washington. Though certainly a strange sight – A bicyclist with an antenna-sporting helmet riding up to the border – the everyday person's 'strange' is this border crossing officer's 'everyday'. He asks perfunctorily, "Have you any money?" to which I reply, "Not much. Two checks." He then inquires, "Do you have any fruits or vegetables from Canada?" I reply, "No," but immediately afterwards my body language makes it obvious that I was thinking more carefully to provide the correct answer. He could care less. "Welcome to America," uninterestedly ending our interview.

Wednesday, November 16, 2016

I began my day in Maple Lake, Washington. I awake to a pristine mountain lake with arctic birds taking refuge in its perfect calm. The gentleman who is hosting me is the father of a friend. As we drive from Sumas, where he picked me up, he animatedly explains to me how floodplains, or as he called it, "the alluvial fan" are fertilized during intervals of intense flooding, where rivers burst their banks. As they do so, they deliver their payload of nutrient-rich sediment as the water slows down and temporarily settles over the plain. I would later learn that most dams halt this fertility cycle of nature. His personality is wild and eclectic and perfect to set the stage for this journey of fifteen-hundred miles. He has a Mexican lover in Puerto Vallarta (and as I leave, he even invites me to follow him there in a few months). He is a tree surgeon. His seven buildings on his property total no more than a thousand square feet. He owns a bicycle boat, as well as a boat.

He starts me on my journey on the Washington State Road Nine, just north of Deming. Years later, when I visited again with my then-wife, he humorously related a memory of me pedaling away that morning: "I send him on his way, worrying a bit about his safety. He soon enters a *narrow* bridge without as much as a glance over his shoulder. Not within 30 seconds of letting him go I see a massive lumber truck with a full load pass him with no less than two feet of distance and him not bat an eyelash. I thought, 'Oh Lord, this is not a good idea!'"

He leaves me with a hunk of Swiss cheese, a jar of peanut butter, some worn but warm gloves, and a small amount of cash for me to spend at Everybody's Store. I have the special opportunity to meet Jim, the owner of Everybody's, the first health-food store in Whatcom County, Washington, since 1970. It's got a little bit of everything, including an expansive meat and cheese selection. Some items are pricey, but a lot of them are affordable and the product of their backyard bounty. Their produce section is nearly all supplied by the few-acre garden behind the store. I take the opportunity to talk with Jim briefly. As we stroll along in the perfect autumn sunshine, the valley looking beautiful – like the idyllic images you might see on organic food boxes, except this is totally real. We talk about healthy soil, plants that pollinators love (such as borage, henbit, and mints), and a bit about my cause. I see that a great first step to protecting our pollinators is to support such stores like this. Compensating the honest earth-steward is the first step toward becoming stewards ourselves.

Thursday, November 17, 2016

After spending a peculiar but lovely night and morning with my uncle in Bothell, Washington, I head to Seattle. After Vancouver, I was not terribly interested in spending too much time in a big city. I believe I only spent four hours in Seattle. After a bus ride into town, the Space Needle is my first destination. As I dine on fine local cuisine, the city is in full view from 300 feet in the sky. The dining room completes a full rotation every 47 minutes.

After seeing things at the macro-level, I zoom in for a micro-experience at the town farmer's market. Among America's farmers' markets, Pike Place is a famous one, and I see, hear, and smell why. From street musicians to grand produce and seafood stands (many offering samples), this is the best I've seen in local fare. Before I lose my head and spend all my money, I catch the next ferry to Bremerton to spend time with an old friend to fill my need for rest and human company before the next wilderness portion of my trip begins. The three days following I spent in Port Angeles, Washington where I met Mark.

Monday, November 21, 2016

Sunshine pervades the entire blessed day. A large portion of the day was spent with the winter sun hanging just below the ridge line. Crescent Lake, lying to the north of the Olympic Mountains, offers the first glimpses of the mighty trees that will be seen in the days to come. In the darkest corridor of the Olympic Peninsula, I revel in a chilly paradise of rugged seven-foot-diameter trees (some measuring as much as ten), damp moss and ferns – ranging from the miniature epiphytes that live in the air on trees to the majestic spreading ground ferns reaching as tall as eight feet – which rarely fade with what is called the dry season. These shadowed segments drink deeply of the wells of moisture from detritus laden with water and coniferous aroma. These ancient trees hold millions of mosses, lichens, ferns, even tiny saplings, so that the largest trees were a forest unto themselves.

As I emerge from the mountain's shadow heading west, following the sun's course, her beams shine thru the trees with their green dangling moss luminescing like the folds of a heavenly white gown. The short branches of each tree shine in their perimeter like the arms of a holy angel, clothed in linen, declaring the great mystery of life. The whole of the landscape, forested in both trees and lichens, takes on an overwhelming hue of yellow-green, as if rejoicing in the day's first sun-food. "This is life at its pinnacle," I think reverently. Later in the all-too-short day, the Hoh River accompanies me as I meander wearily to my night's destination. Its flow is a peaceful end to a most glorious day.

Tuesday, November 22, 2016

A wonderful day of botanical discovery. I personally measure a few trees via my personal method. Knowing that my arm-span roughly equals my height, I give the tree an unembarrassed hug, making note of where my fingertips are. I repeat this in succession until I have circum-navigated the tree. I then reverse-calculate the diameter by taking the measurement I have found and dividing it by three (an approximation of pi). One comes in at a whopping diameter of fourteen feet: the first tree I've experienced of that gargantuan girth. Nothing, however, could have prepared me for what I'd see my last hour in Olympic National Forest: the Quinault Giant Sitka Spruce.

Being in the presence of that tree is a holy experience. As it is over a thousand years old, and an astounding 17 feet wide at its base, touching the bark of the actual trunk requires an eight-foot *climb* up the slippery tops of the root structure. Forests comprised of trees of this size begin straining themselves heavenward as their roots, often swollen to the size of trees, push down into the earth. Trees of this rank reach in both directions in an attempt to connect Heaven and Earth. Verily, Old Sitka did such for me that day.

It beggars the imagination that such forests of 25-30-foot diameter trees could exist, but they indeed have. Old Sitka was once not a lone spectacle, but one of a grove of *hundreds* of trees of similar size, some even surpassing its grandeur. Our hints towards forests of the afore-mentioned girth are seen in the dead though still-remaining bodies of trees that fell hundreds of years ago. There are many trees that claim superlatives such as "oldest", "widest trunk" and "largest canopy", but in my mind, only one tree can claim the title of "greatest." That is General Sherman of Sequoia National Park. It occupies a remarkable 52,513 cubic feet, is around 2.5 millennia in age, and has a base-diameter of 25 feet. The Crannell Creek Giant, a coast redwood, was around twenty percent larger in volume than General Sherman. This volume is over-shadowed by a still bigger woody organism: the Lindsey Creek Tree. This behemoth was estimated to be nearly 90,000 cubic feet and likely surpassing a 30-foot trunk diameter at its base. I am humbled and in awe of these facts.

Wednesday, November 23, 2016

The afternoon of the 22nd mercifully brings a happy-go-lucky stranger into my life: a Salvadoran middle-aged man from Los Angeles, cycling through the same country as I. I loved his words and open-hearted nature so familiar to me as one who has bicycle toured Califor-nia and lived there for years in the past. "I saw you put your bicycle on the rack and thought, 'Oh wow, there's someone doing this who is as crazy as me!'" We share a hotel and a meal at Denny's, dry off our clothes and work on our bicycles, merrily conversing all the while.

We don't ride together long, as our paces and tolerance for the cold are quite different, but we keep in touch. As for me, my destination is Astoria, just across the Columbia River in Oregon. We part in

Raymond, and I press on, determined to make the border. The shortest route takes me a few miles east of the Astoria Bridge, so I get a taste for the wind I am going to contend with.

This portal into Oregon is a four-mile long, narrow-necked monster into which I plunge without a second thought. Almost instantly, winter screams across my ears with her 40-mph winds, accentuated by the bridge structure. It was so difficult that the rain actually stings my face. So difficult that I must steer to the right to keep the wind from driving me into the traffic lane. When I reach the final bridge section, nearly 200' in the sky, the elements seem to conspire against me. These torrents and gusts cared not for my existence. The rain clouds seem to form endlessly along this trussed section of bridge, always providing a stinging swarm of raindrops to accompany the howling winds.

At one point, nearing the top, pressing forward is simply impossible and the only thing to do is perch my right foot upon the small concrete girder and hope the wind dies down enough to proceed. The sour taste of dread momentarily gathers in the back of my mouth. The moment passes, along with the impenetrable wind, and as soon as the road sloped down, I pedal with all my might. As the road begins its 270-degree curve towards ground, I look to the side almost in disbelief that I had crossed it. Though it had passed and not devoured me, seeing it afar off aroused something akin to terror and awe in my breast, looking ten times more awful than before. But I had made it, and Portland awaited.

6

The City of Roses

After the most nerve-wracking bicycle ride of my life, I steal away by bus to the lovely Portland, a place that feels to me like a foreign country, rich with sharing and diversity. Due to arriving at 8:30 at night, a hostel is my best option. This is essentially a boarding house with all the amenities of a hotel, except cooking space, bathrooms, and sleeping quarters are shared. I quite adore the set-up, and though comfort was offered through our plenteous accommodations, my mind races with the adrenaline of the Astoria Bridge, thoughts of what lie ahead in Bakersfield, California, and the current surreal-ness of my life. Sleep eventually comes, and the morning stirrings of my temporary housemates gets me up and ready to set up for the day.

A remarkable positive spirit drifts about me as I seek out a location to hang my hammock. Strange to say, this body has melded itself to the hammock more than the bed, just as it has long since become one with a bicycle. Now, traveling without them leaves me feeling spiritually and physically lacking. Due to previous pilfering of my belongings, over-ambition leads me to a steep and obscure road, to a nature-trail, and finally a two-minute climb up a steep jungle-like hillside where chances of being discovered are virtually zero. The rain is almost a constant companion, and as I set up the tarp, struggling to make the hammock hang well on the slick, lichen-laden trees, Van Morrison's *And it Stoned Me* prances through the corridors of my happy and tired brain, emphasizing the words, "Ohhh the water. Hope it don't rain all day."

Later that Thanksgiving Day (back at the hostel) finds me gathered around the residents of thirteen different sovereign nations, including Iraq, Malaysia, Mexico, and Czech Republic. It was a powerful renewal of feeling to be a citizen of the Earth while surrounded by these colorful and diverse people, as nearly all religious, socio-economic, or political/state entities have eluded my allegiance. Now time for dinner! Everyone is encouraged to either offer a donation or a dish for dinner. Thus, a bounty of foods from many cultures fills us both spiritually and physically.

My last day is an absolute rush. In the morning, I take a bicycle ride with the bishop of the local Latter-Day Saints congregation and his friend along the bicycle trails in southeast Portland. It is comforting to be guided in unfamiliar territory. He even helps me discover some cultural gems during our travels. One section of neighborhood connecting separate bicycle trails has a beautiful intersection with painted murals, free herbal tea for weary travelers, and small huts designed from clay and straw and other raw natural materials, from which gracious producers sell fruits, vegetables, and crafts. A large building across from one lagoon sports a mural covering its entire visible face with a backdrop of sky-blue, displaying the birds that pass through and inhabit the bird sanctuary through which we cycle. Many of the featured species are absent, heading south for the winter. I will soon follow their path, perhaps even to balmy Puerto Vallarta, Mexico.

My next destination is a holistic-healthcare school, where students can be taught whole-foods-based nutrition, Alma Therapy, Qi Gong, other Eastern healing practices, and more. It is wonderful to meet with one of the directors, as nutrition has been one of the most empowering life-skills for me, and I seek to empower others as well once those skills have been sharpened and directed. After my travels, I later discovered the philosophy of Dr. Natasha Campbell regarding the healing of the gut and consequential clearing of many mental challenges (from the more-common such as depression and addiction to the occasional such as autism) and have since devoted a lot of my health-promotion efforts towards her program.

The famous lawyer's ride in Downtown Portland is my next destination. Twenty-five years ago, two lawyers began taking their hour lunch break to stretch their legs and see the town via bicycle. Soon, many other lawyers joined their weekly promenade. For more than a decade now, anybody is welcome and come rain, sleet, or snow, at least a few people have gathered weekly for this ride on the southwest slopes of Portland.

We begin with a long, slow ascent of SW Broadway up to Portland Heights. We then carry onto SW Patton and all-the-while I'm thinking, "When are we going to stop climbing?" A couple of cyclists kindly stay behind to ride with me, though I determine that they too can't compete with the strongest cyclists in the ride and thus aren't falling back *too much* on purpose. An important detail to call into attention is the fact that I was the only one who was on a bicycle tour and thus had to carry some twenty-five to thirty pounds behind me. Another rider notes this: "I'm pretty impressed you're keeping up with that weight. You're a strong rider." Inwardly I beam at that comment and continue to push hard.

I haven't pushed like this since I raced in Tour de Franklin (A 50-mile race in Macon County, North Carolina, now discontinued), where lungs burn, muscles heave in a great flourish that they secretly love, and sweaty palms can barely keep on the handlebars for the side-to-side motion of your cadence, giving it all you've got. It hurts so good. Add to this the winter detritus combined with the slight rains resulting in questionable traction, and the downward sections are exhilarating.

This is not just for the thrill of gravity's pushing, but for dancing on the line between taking the mountain-hugging road too fast or too slow. Too fast, and inertia wins, flinging you into ditch, rock, or Portland below. Too slow, and friction wins – you don't have any fun! For a few glorious minutes, as we pass the sunny side of the ridge, Portland comes into view, basking in the wet sunshine. The roads steam, smelling like Pacific Northwest moist fall, and glistening trees break up the cityscape below with their wild beauty.

Another remarkable meal occurs in the evening. Food Not Bombs puts on a lavish vegan meal twice a week in northeast Portland, simply because. The feast occurs in a park on 18th and Stark. The food is prepared a couple blocks away and is transported the short distance via heavy-duty bicycle trailers. Some can carry up to 500 pounds! Local food banks, gleaning agencies, and charities supplement their provision of the food. I am blessed to meet so many good people, and the goodness keeps on coming. A kind soul summed up the organizers' motivation in his parting handshake and words to me: "Thank you for being you."

I think we could all learn a little bit from the people who contribute to this unique place. Pollination functions on the biological principle that sharing information (such as genes and hormones) and resources among species creates resilience and adaptability in ecosystems. Bees, butterflies, and other pollinators ensure the future of their food source by mixing up the gene pool of trees, bushes, and annuals that provide them with nectar and pollen. Pollination provides genetic diversity as well as reproduction services. Their dance from flower to flower provides this interchange in a world ever subject to change. The more diversity of ideas and positive passions, and the more intermingling of those through proximity, open dialogue, noncensorship, and sharing, the more we might succeed as mankind.

. . . Couldn't we learn something from the humble bee?

7

Snapshots: Oregon at Land's End

Tuesday, November 29, 2016

My night at Cape Lookout campground was awash in spiritual feeling. I walk the beach in the alive night and am surrounded by two oceans: to the west, a terrestrial; to the east, a celestial. The clouds above encompass me as they move in, covering my sea of stars and providing a fresh soak. The morning holds more beauty. Within five miles of my departure, I see the entire curvature of the bay from whence I came. I behold it all for a great while. Stone islands lie upon foundations of evanescent mist. The dunes viewed from Gammon's Launch seem just as fleeting as the fog that flows in, only to both be washed away in the blink of an eye. In this moment, my life feels just as transitory and beautiful.

I see the name Dick Gammon upon a headstone: "Died 2009 Pioneering hang-glider pilot." I imagine him launching from this spot out into the blue, his last footstep perhaps touching his would-be headstone, the summer wind at his head providing lift. Did he die in this very hour, with this view in his eyes? Did he live for Earth's beauty?

And ultimately succumb to her boundless power, hands lying upon an aviary altar? Eventually I am released from the trance of looking out on this scene. Not five miles later, I perhaps have my answer, but with hands upon bicycle handlebars rather than a hang-glider, and though I did not perish. The sun begins to break as I reach the top of Cape Lookout Hill, the drops of liquid and shine pouring down, mingling together onto the churning ocean's surface –An intense brightness. I keep going instead of stopping as before, for it truly seems that if I were to gaze long enough, it's sheer beauty would be too much to bear, and I be torn to shreds in its transfiguring light.

"True, there are innumerable places where the careless step will be the last step; and a rock falling from the cliffs may crush without warning like lightning from the sky; but what then? Accidents in the mountains are less common than in the lowlands, and these mountain mansions are decent, delightful, even divine, places to die in, compared with the doleful chambers of civilization. Few places in this world are more dangerous than home. Fear not, therefore, to try the mountain-passes. They will kill care, save you from deadly apathy, set you free, and call forth every faculty into vigorous, enthusiastic action. Even the sick should try these so-called dangerous passes, because for every unfortunate they kill, they cure a thousand."

- John Muir, *The Mountains of California*

Thursday, December 1, 2016

A singular experience greets me in Port Orford, a little town complete with a generous library and a cute health food store. Feeling strongly decided upon Port Orford Heads despite it being impromptu, I laboriously ascend the 200 foot-climb road as the sun flees from the sky. The sound of surf is omnipresent – a quiet sentinel of the black night. I scout out the whole area, going out to the point where the sea to the south is in full view, the last hints of dusk visible on the horizon and tracing inward to the rocks by concentric grey lines of foam. Another vista, looking to the sea west, gives a nice view and feeling, but is too open so I decline setting camp there.

A feeling of uncertainty arises as I return to the street. Returning to the vistas from the parking lot, a certain impression not to set up to the right of the path comes to me. Moments later, I shine my lonely white lamp to the west and a set of metallic blue eyes meet mine. Almost at once the thoughts, 'nocturnal animal,' 'eyes too far apart to be,' And 'big cat' race through my mind. Gripped in this ocular exchange, we have a moment of tensile silence. A chill runs through me, like shivers at first, then solidifies into a nearly tangible sensation in my soul, real as steel on bare flesh. Beauty and terror mix as I feel the raw essence of this animal: curious, fearless, and potentially deadly. I break away from this gaze and ride swiftly to where path meets ruthless rock "heads" tumbling into the ocean's foaming embrace. I suddenly feel protected, the sea ever my rock and stay.

"Walking. I am listening to a deeper way. Suddenly, all my ancestors are behind me. Be still, they say. Watch and listen. You are the result of the love of thousands."
- Linda Hogan, Chickasaw Storyteller

My prayers for protection were answered in a beautiful way. A calm assurance is felt, that somehow whatever threat laid in the brush could not reach me here. I think about the ancestors I discovered in my research and felt that they were watching over me, guiding me, even protecting me from wild beasts. The following morning finds me hanging happily between two oaks overlooking blue infinity, the perfect wind rocking my hammock.

Friday, December 2, 2016

Though the serene splendor afforded me at Port Orford Heads was hard to leave, I continue onwards. Humbug Mountain State Park, with its 1700-foot-tall peak just one-half mile from the sea, doesn't provide a seaside hiking experience, but rather chills me, covering the winter sun from the south. After a quick and cold shower in the State Park, I continue. There isn't a terrible amount of rock features past the Three Sisters, though the omnipresent beach and surf accompany me and provide endless opportunities for sun-drenched relaxation. I take many such opportunities, often only needing to step 10 feet away from the pavement to lay in the warming, white sand.

A lovely surprise awaits me: a young bicycle-tourist couple from Austria. I have overestimated the number of companions I'd find during this pacific-coast trip. Only they and Mario have ridden beside me, and for only a short while. This time, I am the one left behind. I find myself having difficulty keeping up over the ten or twelve miles before Gold Beach. The tunnel of vegetation on a bicycle trail we follow creates a feeling of smallness, which produces a counterpoint of enormity when it opens back to the main road just before the bridge over the Rogue River, which river is nearly 1500 feet wide at the point of crossing. They go on their way, perhaps to Brookings, and I stay in what would become three days in the Gold Beach area.

Saturday, December 3, 2016

Much of the time before Sunday is spent roaming around town, walking the beach, and working at the library on the accident claim from being struck by the vehicle in New York. A few large trees lie on the shore of the beach, and upon mention of it, one local explains that during a large storm a few years back, redwood trunks rolled onto the beach for many months before rolling back into the surf, never to return.

Saturday evening, I meet a fascinating character. He also carries the spirit of a wanderer but is actually a local who tells an amazing story on how he got here: "Well, my wife died years ago, and with my parents both passed and gone I decided to make a new start. I had no idea where else I wanted to live. So, I literally took a United States map and threw a few darts behind my back, and I read 'Gold Beach' underneath the first dart. I didn't know anything about the place, and had never even been to Oregon, but figured that is where I was supposed to be." The excitement and expression in his voice is awe-inspiring, as I can tell he truly is taking this new environment on as if it were handed to him by the universe, as if he were directed there by God. It is often cloudy and gloomy and only a family at church opening their home to me for much of the time keeps things jovial and pleasant.

Tuesday, December 6, 2016

The corridor between Gold Beach and Brookings where land ends was marvelous in my eyes. As the journey continued south, fatigue seemed to take its toll. 60-mile days turn to 40 and 30. Eventually twenty miles is all it takes to wear on me but exposes a glint of gold in my soul. As I moved through the 'banana belt', a microclimate between Port Orford and Brookings, Oregon, temperatures warmed five to ten degrees. Combined with the transition from the previous days' clouds to sunshine, it felt like an emergence into paradise. Points like Arch Rock, Natural Bridges, and Thomas Point received my full attention. I even napped upon the rocks forming one of the bridges during the gentlest rain.

Precipitation and filtered-to-full sunshine dazzled my soul the last half of the day. Though the bridges lay hundreds of feet below the viewpoint, I couldn't resist. While upon them, I shall never forget the sight received in my sleepy eyes of sea stacks alternating to left and right of view, both fading in detail and brightening in the distance as sunlight poured in stronger through thinning clouds. All was enveloped in a veil of rain dancing in the glorious day. The sea chanted and churned in reply, joining in happily with this song of water and light. I even received a shower of hail for an enraptured minute upon these bridges.

Later, this same sea would fill me with irrepressible joy, overflowing with gaiety. While sitting in meditation with the sun now at full strength upon the sheer cliff which I laid, the sea some 350 feet below sent surging foam up against the rocks with calm repetition. After a time of contemplation, allowing all this to flow thru me, it began to resonate in the soul. I then knew the love-power in those foamy crests, beating upon the land in endless virtue. The ocean and land were one with me and I laughed. For more than a minute, deep belly laughs came over me and hands clenched thighs and grass. I laughed because of nothing; it simply came. . . Inexplicable happiness. Clean and innocent as a sunbeam was I – a vessel of the Earth's joy.

Oregon's end seemed to recede forever from my grasp with each mile, as my muscles were waning along with the day. Once in California, I detoured on Ocean View Road to take in the last bit of pink-orange ocean-side splendor as well as avoid the busy traffic which is most dangerous to bicycle alongside during dusk. As I wearily cycled through the last dozen miles before Crescent City, I came to the border of Redwood Country. Though their scale was impossible to appreciate in the night, I felt their enormous presence as their moon-shadows filled the roadside.

I took shelter in Jedediah Smith National Forest. A feeling of ancientness and stillness surrounded me as I slept – almost so peaceful and quiet I struggled to drift off – and as I awoke, the largeness of the trees took their full effect on me. I simply stared up into the heavens for more than an hour as these cathedral pillars, which seemed to hold up the sky, towered silently over me. Eventually, I departed and headed to Crescent City, where a kind churchwoman awaited this wayfaring stranger to put him up in a church for the night. She showed me books full of notes and signatures from passersby, including a group of signatures and notes from a troupe of a dozen cyclists. The town struck me as *the* perfect place to live, a humble sea-side villa. It felt as old, as knowable, and as uncaring of time as the redwoods. It seemed capable of holding this wanderer not merely for a day, but for a peaceable lifetime.

Thursday, December 8, 2016

The morning began early. Preparations for a bus ride through rainy, foggy weather had me scrambling. As excellent as the nature would have been to bicycle through, I had seen my share of redwoods and would see the Lady Bird Johnson Grove during my stay with a friend in Northern California, and the fog and rain would have obscured much of it and lent too much danger to the ride. My friend Tim Connell was awaiting my visit and I was looking forward to a few days to recuperate. After the bus dropped me off, and a half an hour of wandering, I finally found his place.

It was a wonderful time. Tim took me out for breakfast once I got unpacked, and later we explored Gold Bluff Beach and hiked Fern Canyon, which follows a shallow canyon river. For some reason, I insisted on doing it barefoot. The water was cold and the river topography was unforgiving of my weakened body. Later that day, I think hard, as I rested, about a conversation I had with a well-seasoned bicycle tourist in Brookings. He informed me that the amount of muscle tissue breakdown and rebuilding that happens on long, multi-week rides is immense. After asking me about my diet, he assured me I was running a severe protein deficit. I couldn't help but feel I had cheated myself of a better ride. Luckily, the travels were but half over and after much rest and nourishment at Tim's house I could hopefully start out the second portion of my travels on the right foot.

8

—

East Meets West

"Permaculture isn't always about feeding ourselves. It's also about growing and being part of the nature around you and attracting beneficials into your area. So, you don't just plant things you can eat. You plant things that are good for those beings you want in your area"

-Tim Connell (2016)

As I complete my travels for this portion of the journey, I look forward to the beautiful, dry home of an old friend, Tim Connell. We met years ago when I lived in California and quickly became friends. At the time, I was just beginning to learn about natural lifestyles and often inquired about his habits to learn more. When he moved to Northern California, I determined I would catch up with him again, and here I am. I feel very comfortable talking with him as an old friend and add many of my own thoughts. So you might consider this a conversation more than an interview.

His abode rests in the quiet east end of a lagoon, offering excellent views of the Pacific and the lagoon below. The traffic on the U.S. Route 101 is naught but a peaceful whisper by the time it reaches Tim's back door, whether to his barn or home. He loves the moisture that blesses the trees and variety of wild edibles, which are present even during the wintertime.

Tim Connell is a pharmacist and a blooming permaculture practitioner in Coastal California. I am curious how Tim was motivated to practice pharmacology. "When did that become a passion or motivation for you?" Tim frankly replies, "Well, my motivation was for my daughter becoming a diabetic at age five, and my inability to support myself as a grade-school teacher." "How much schooling did that involve?" Tim answers, "Well unfortunately, too much. I've got 12 years of post-secondary education, but it takes a minimum of 8 years to become a Doctor of Pharmacy." Trying to get the full picture, I ask about his role in conjunction with medical doctors. He explains, "Well, I am the last gate to making sure that whatever is entering the patients' hands is not going to harm them and that it will not interact with the disease state or the medications they're taking. Ensuring that the right thing goes into the bottle and that the right directions are written on the bottle."

Having parents who are in their seventies, I am relatively familiar with the complexity of pharmacopeia. "I know that so many things can go wrong with drugs, which can be very similar-sounding and similar-looking. The wrong label can get attached, so the devil is in the details, right?" Tim makes an excellent cross-industry comparison. "I had an engineering friend, and he says their quality assurance is to check every tenth part. But you can't do that as a pharmacist, you must make sure *everything* that goes out is correct. You'd fail if you only checked every tenth prescription. It's monotonous and mind-racking, so only certain individuals are good at it. My staff pharmacist who I've hired is amazing at her attention to detail. She has not made a mistake in the three years she has worked for me. She has caught my mistakes, and I don't do that many. Thank goodness I haven't made any major mistakes. I've made minor mistakes such as quantity of pills in the bottle."

Switching gears to learn more about his other passion, I ask, "Now, how did you become interested in permaculture? What, if any, role did your father play as a famous biologist?"

Tim feels a trace of irony in my question and jests, "Well, *that's* a loaded question." We laugh and he continues. "I became interested mainly through my son, who is an organic farmer. He introduced me to the concept and gave me a few names to follow such as Bill Mollison. *Gaia's Garden,* by Toby Hemenway, was the first permaculture text I read, which was an inspiring book – Good graphics and lots of detail.

"I also built a straw-bale home in Arizona with the thought that it was a renewable and locally available resource, and this was in the early 90s before I had even heard of the concept of permaculture. I guess that's where I began thinking along the lines of 'renewable', 'carbon footprint', being 'kind to the planet'. And also things that don't require deforestation and things readily available in your area: dirt, rammed earth, cobb, and straw which has been slash-and-burned for so long.

"My son's surrounding network introduced me to Permalytes, Mark Shepherd, Jeff Lawton, Joel Salatin. The state of our economy and the way people live nowadays also drives my interest." Again, thinking of my parent's health issues, I shudder a bit and state, "And as a medical professional, you've seen the worst of it. You've seen it all." "Yeah. Becoming aware of our gut flora is important too."

Tim turns to talking about his father, Joseph H. Connell. "My father taught me critical thinking, questioning everything, loving knowledge, but not trusting all the written words and what you hear. I learned a lot from my father-in-law about love in business aspects, such as loving those you work with, knowing how much to trust people with, knowing when to reward and when not to, and learning how to get what you want out of life without hurting another individual. He was a very good mentor." I really feel what Tim is saying and expound, "Others don't have to lose for us to win. So, it's nice to hear your thoughts about living equitably. One of the things I've learned in permaculture is the spirit of cooperation. Though there is some element of competition, all elements in nature are working together in harmony."

I ask a general question and Tim replies with a wonderful definition of and an explanation about the ethics of permaculture. "So how have you seen permaculture practices benefit pollinators in your region?" "Well, permaculture isn't always about feeding ourselves. It's also about growing and being part of the nature around you and attracting bene-ficials into your area. So, you don't just plant things you can eat. You plant things that are good for those beings you want in your area. You also plant in abundance so that things you might see as pests do not take away from your inner sphere. You give the deer plenty of cover crop on the periphery that it fills them up so that they don't have to come into your food and cash crops. The philosophy there is that if you can't beat 'em, join 'em'."

I am intrigued and ask about the work of the local permaculture guild. Tim explains, "They do scion exchanges. This is a twig or root of a plant that will be grafted onto another tree. Fruit trees often need a parent's plant segment to grow from rather than seed to produce a consistent product. They also have an extensive seed exchange program to create more variety that is adapted to the region, and this is all open to the public. They do this to get an heirloom species of familiar foods developed in the area."

I excitedly add, "To create some local resilience and increase the bio-diversity." Tim affirms and I continue, "That's really cool. I think one of the inputs (a product sourced non-locally) that leaves one most vulner-able to rely upon is seeds. I was talking with a Port Angeles beekeeper, and one of his concerns is the lack of local queen-raising in beekeeping in general. There's a bee-keeping unit in Washington State that only works on queens. This group is raising queens in Washington to try to get them adapted to that maritime climate. If you raise queen bees and California and try to get them to perform in the cool maritime climate of Washington, they're likely not going to do so well. It's the same thing with the seed saving. People may say, 'well, these varieties aren't doing well up here'. Well, they *could* if they become adapted to the climate in their genetics (over multiple years of planting and harvesting seed)." Thinking of what Mark Shepard, a famous permaculture practitioner and owner of New Forest Farm in southern Wisconsin, has said in this regard, he concludes, "Like Shepard has said: Plant in abundance and whatever doesn't survive, 'well good riddance.'" "It wasn't meant to be." Tim concurs, "Right, wasn't meant to be."

I am loath to close, as there such crucial information to be learned from this man about pollinators and engendering general biodiversity. I ask, to finish up, "What are some recent things you have been learning?" Tim replies, "I'm learning a lot about gut flora, and how we need a diverse diet of nutrient dense food." I ask further, with the bees as a consistent context, "As someone studying permaculture, you get opened up to a mindset that is starkly different from our current economic mentality and values. What are some of your recommendations for people who want to shift their mindset to partnering with nature, especially for the sake of our bees and other pollinators?"

I couldn't have said it better myself: "We need to stop forcing nature to conform to us and instead conform to her, so we will eat what nature gives instead of forcing her to produce what we want. That includes pollinators. We need the food that they help produce. Plant something besides grass, things that flower, things that are native and that can make it without you there and feed the soil. We need to capture the water and grow multilayered soils. I totally believe that we need to get people out of the cities and get them back into the small towns, developing those skills of production." Enlightened, I conclude,

"So, we need to grow those healthy resilient communities?" "Yes, just like we're growing the soil."

There are few things I've heard during this journey that were as potent as what Tim offered in the last paragraph. During and after World War Two, America kicked its industrialization into overdrive, seeming (in hindsight) to get as many people off the land and into non-land-based occupations as was possible. This was "progress". This was "lifting the people out of poverty". But it turns out that the land has suffered from humans disconnecting from their land-base. The ecology has atrophied as direct result of the people morphing from caretakers of the living land into disconnected consumers of its resources.

This 're-grarian' ethic Tim touches on – essentially to become agrarian again, to return to a land-based consortium of local, 'cottage-industry scale' economies -- is the reversal of this trend described above. This involves re-teaching many to feed the bees through bio-diverse plantings as well as directly keeping bees of many breeds and species.

The conversation went on after this point, discussing the farming success of his son, the non-necessity of landownership for farming, as well as the idea of improving our own situation via ingenuity and social capital instead of waiting for prescribed requirements dictated to us by a money-preoccupied society. Specifically, we talked about Curtis Stone, Rob Greenfield, and others who have made a decent living by raising produce on land they don't own. Many would enjoy adding value to their community through hosting gardens and receiving free produce and are only waiting for the asker. If the definition of success ought to be diversified – not only in terms of money – then so too should the definition of cost be diversified. It was a great blessing to see Tim again and look forward to keeping in touch with him as the years pass.

9

Honeybee Haven

"This is no country for bees," I consider as I put my feet to the ever-gracious pedals of my "Vitamin B" road bicycle, finishing up the second portion of my travels. My location: Davis, California. I had come from the idyllic wine country of Santa Rosa and Napa. The visual and botanical harmony that arose from grapevines, wild mustard, and other lush wild growth in the fields stood in stark contrast to that which lies before me.

I pass through almond orchards and empty fields, all sectioned off by seemingly unworn blacktop roads. It has always been an other-worldly experience to bicycle these corridors of agriculture. It is unlike anything that could ever be found in nature: right-angles, straight lines everywhere one looks, and no diversity of life whatsoever. Cold silt-clay, quiet barren trees . . . The whole world about seems dormant, only waiting to come to life come spring, where these trees' survival would then be possible only through the power of irrigation, imported honeybee hives, insecticides, and fungicides.

Arriving in University of California Davis, with its curving roads and diverse botanical gardens along the campus river, feels like an oasis. After passing lovely entomological murals and info-boards, the western portion of campus, and a final stretch of fallow fields, I come upon the H. Laidlaw Jr. Bee Research Lab. During this 60-degree Central Valley winter day, there seems to be no warmer welcome than this research facility and the adjacent Honeybee Haven which is sponsored by Haagen-Dazs, an ice-cream company. I am greeted by Charley Nye, the facilities' lab manager. His energy and youthful openness have me feeling optimistic about my hours here, and this would be affirmed once he showed me the Honeybee Haven.

First was an interview with Dr. Robbin Thorp. Dr. Thorp (now passed) is an emeritus professor of UC Davis at the H. Laidlaw Jr. Bee Research Lab. His work with honeybees, native bees, and his uncanny ability to identify wild species has won him widespread acclaim. A current focus of his is tracking the Franklin's Bumblebee in Northern California and Southern Oregon (including the banana-belt I had just traveled through).

I start out by asking Dr. Thorp how he became interested in bees. He answers, "Well, not until my senior year of college. I took an entomology course and got interested in insects and I asked the professor what a good project would be to do at the end of the year, and he suggested bees and the pollen that they collect in springtime. So, I spent the winter learning to recognize different species of native bees and learning about pollination and pollen morphology. That just got me hooked. I started getting into literature and just got fascinated with the pollination story, the interaction between bees and flowers. I've been at it ever since." Hearing his story brings back warm memories of gardening and I relate, "I had a garden I got to play with for a few seasons and I have to tell you, there's nothing more gratifying than planting that clover, or that sunflower, or whatever it is you plant, and watching the pollinators come."

Dr. Thorp relates the idea to his situation at the lab. "The garden we have here is very therapeutic for me. After sitting at the microscope for hours at a time, I just have to push myself away and go for a walk in the garden. It's not a brisk walk, it's not exercise, but it gets me moving and watching what the bees are doing. There are so many interesting things going on out there."

Thinking about the centerpiece of this visit, I excitedly ask, "So, in that bee haven, how many pollinators have been seen out there? Who did that count?" He smirks and proudly answers, "I started that. I started the year before the garden went in. I found forty species before the garden even went in. The next year when the garden had been planted, we found forty species, but only twenty of them overlapped, so now we're up to sixty. And now the tally is over eighty species of native bees that have been found over time. Not all are common, but all are present. That's a pretty fair diversity of bees that can be found in a pretty small garden, and for being overwhelmed by honeybees in the apiary nearby." I am thrilled by his telling of the diversity found therein, and he adds that managed bees and wild bees indeed *can* co-exist in the same space.

> Professor Thorp has a long history in the Central Valley and has seen a lot of changes. His keen eye has been fixed on the bee all the while.

I ask Dr. Thorp, calling upon his 50-plus years of agriculture experience in the central valley, "How have you seen the land use changes affect native pollinators in the area?" He explains, "The honeybee and native bees have been dramatically affected since I started. Since I started in the mid-sixties, the almond-bearing acreage in California was 90,000 acres. Now that number is over 900,000 acres. At two colonies an acre . . . Colonies have come from all over the country out here. That has impacted the bee industry tremendously. Irrigation of crops in the summertime has opened up areas in the Central Valley for native bees that live in the foothills but can live in riparian areas and areas that retain moisture. The bee fauna has persisted and expanded in the area. Even some of the weedy invasive species like yellow-star thistle comes into bloom at a time when most of our native flowers are going out of bloom. This has extended the life cycle of some of our native bees and caused a lot of their populations to increase, even though there are many efforts by cattlemen to eliminate it. It produces marvelous honey."

I am surprised by this information, and check my understanding, "So, you've seen the irrigation bringing in *more* bees?" Dr. Thorp sorts out my understanding and explains, "Well, of course a lot of the agricultural intensity has reduced populations by having taken out land that was formerly occupied by some of these bees, so it's restricted them and changed them. Particularly where pesticides have been used. That's not very conducive for bee survival. So, there have been both plusses and minuses. But a lot of the bees still persist. There's a lot of bees in urban areas.

"I have a colleague who is looking at bee population in fifteen urban areas throughout the state of California. We find on average upwards of 100 different species of bees per area. That's a pretty fair diversity." I am encouraged by this common theme that bees seem to need only relatively small boosts in resources to succeed. "It seems to me that bees need relatively little space. A little garden goes a long way. I have studied what's called urban agriculture, where a surprising amount of food is produced in some of these urban areas. In many inner cities in non-modernized countries, all their honey that they consume is being produced in-city, and it's just people planting bee-gardens. They like the flowers, and bees like them too. That's really encouraging."

Dr. Thorp affirms and continues, "For honeybees it's a pretty good situation because you have this diversity of gardens well within their flight range, and you have sources of water, so they have pretty much what they need. You've got holes in the walls of houses and buildings where they can nest. Both native bees and honeybees do well in these native areas. It depends on the species, but there's still a lot of diversity there." He positively posits that urban gardens can contribute to keeping bee populations healthy, both in terms of native bees and honeybees.

> Dr. Thorp offers some great advice for those looking to help the bees along

As thoughts invariably turn to how we might help the pollinators, I state and then ask, "On my website I've documented so many people saying that we need to be planting bee-friendly plants, and I'm actually going to be talking a lot with Charley Nye about that. What are some recommendations for resources for people to learn how to plant a bee garden, and to know what species can attract bees in their area?" His energy picks up and tells, "There's actually been a lot of interest in this. There are regional lists, so that people from different regions can get informed on their areas. The best website to get started on is the Xerces Society. It's an insect conservation society that has really gotten into pollinators in recent years. There's a lot of information there. There are books on bee gardening."

Dr. Thorp mentions one of his team's latest publications. "It's called *California Bees and Blooms*. It has broader applications, although most of the plants are selected for here. The principles are the same and bee groups that are profiled occur throughout the country. There are books in the east on bee gardening. There's lots of interest for gardening for pollinators. There are even hummingbird gardening books, and butterfly gardening books."

Comparing the potential of what we could do with how things are faring currently, I ask, "How's your outlook for native bees in this region, with people planting bee gardens?" His demeanor reveals concern but hopes at potential for positive change, "Well, that's a positive aspect of it, but the more intensive agriculture, and the more urbanization, those are the things that tend to cut into pollinator populations. There's an interest in ecological intensification in agriculture, such as planting hedgerows and strips of flowers. Almond growers don't like that extra vegetation in the orchard, because of the way they manage the crop. They'll tolerate crops on the perimeter, but they don't want any flowers competing for bloom when the bees come in." I find this peculiar. It seems so different from what I've come to understand about working with nature to produce nature's bounty, and the puzzlement shows on my face.

He lays the difficult facts before me. "It's kind of the historic mind-set. There are some economic incentives to this newer approach, but it does cost some up front. In California water is a big issue. You've got to be able to convince them that it's a worthwhile investment when they want to put all of their water to their main crop." I offer, "If they see that if they include hedgerows of native flowers, they can soon reduce the amount of money they spend on hives. They then might be able to break out of that mindset wherein everything is input and realize, 'Hey, not everything has to be brought in. Some things just happen if I set up the natural infrastructure'. That's the challenge: Changing mindsets. I hope that as people read my words, it changes their mind."

Dr. Thorp absorbs what I've said and explains how this positive cultural change might occur. "That's another big function of the farm advisers. They are the extension service. They are the ones who get the information from the researchers out to the growers. Education is really key. Your website is that kind of thing. We need to get that information out. If that information is going to be useful it's got to get 'down-level'. Bees seem to get people's attention and are a newsworthy topic."

Dr. Robbin Thorp truly is a wealth of information. After showing him a favorite photograph of mine, he identifies the pollinator as a Skipper. We then talk about our favorite pollinators, his of course including the Franklin's Bumble Bee, while many of mine reside in the Spignidaceae family, commonly called the hawk moths. Many growers of tomatoes, peppers, eggplant, and tobacco revile these creatures in their immature, voracious stage wherein they are known as hornworms, but they do their fair share of life-giving through pollination after their earlier rounds of life-eating!

It is now time for me to go with Charley Nye out into the Honeybee Haven, see some research that is currently going on, and get to know his role in this fantastic research facility. His experiences and perspectives perfectly compliment those received from Dr. Thorp moments ago.

After asking an introduction and how Charley came to work at University California Davis, he replies, "I worked at the University of Illinois Urbana for some time. I worked there for three years, and then I left and was a teacher for a few years. When a job opened to be a lab manager when I was twenty-five, I came back and was at University of Illinois Urbana again for five years. And I've been here at UC Davis for a year now."

He certainly is kept busy at the lab and so I ask, "What experiments are going on currently at the Laidlaw Research Facility?" Charley explains, "For one, those interested in permaculture have been doing some studies. They want to know what bees prefer what flowers. They have this kind of year-round data that gives them an idea of what's in each plot. They can go to a cherry orchard and realize that cherry trees are best pollinated by these native pollinators, and not just honeybees. So, on any spot you cannot plant cherry trees, you choose to plant these four plants that attract these native pollinators so you can increase their population in an area." I am thrilled to hear about a university discussing permaculture, and I explain a potential benefit of nature-based solutions, "A farmer gets interested when an extension tells them these sorts of things: that they can cut down on their inputs and thus overhead."

Charley smiles at the explanation and says, "Exactly. It is interesting. A lot of the people at the Research Facility went to the almond conference last week. In the last ten years there seems to be some change where the farmers are beginning to care. A lot of problems for pollinators is that farming has become overly efficient. [Decades ago] we kind of reached a peak of efficiency. They then had machines to pull up deep rooted trees and plant wall-to-wall in their crop. Now the farmers are starting to realize that maybe they drove away the ground-nesting bees and tore down the trees where the honeybees were, and that they need to do something to help the pollinators." There is a lot of hope in his voice, and I find it infectious. One thing calls for clarification, and so I mention, "I've heard it's hard to get rid of those ground nesters. You have to plow pretty deep." Charley then explains, "But if you plow year after year, it can aggravate them away and reduce their populations."

> Charley sheds some light on all the research that goes on with
> the honeybees here at the facility.

Understanding this and its implications for the imported honeybee industry, I inquire, "What are some roles you fill here at the Laidlaw Research Facility?" "I have to make sure the colonies are ready for studies/experiments. I'm talking with all the labs trying to figure out what their plans are for next year and preparing accordingly. The Niño lab wants to ramp up their colonies for next year because they felt like they want to do larger-scale studies. Each lab runs about fifty colonies.

"In Illinois we ran 100 colonies at a time. For Varroa Mite studies you need a large control group of colonies that aren't having anything done to them and eighty percent of the hives end up dying when winter comes. You're doing a lot of those experiments with the expectation that some are going to fail. We have a huge turn-over of colonies." I am shocked and exclaim, "Eighty percent?!" He matter-of-factly explains this unfortunate news, "If you don't do any Varroa Mite prevention that can be a normally expected die-out rate." Luckily this is the control group, and the managed honeybees don't have nearly this die-out rate, but is still appalling in my mind, especially what I have learned about so many hives faring quite well even without Varroa treatments in areas far from industrialized agriculture.

I observe of honeybees within industrial agriculture, "Varroa seems to be an absolute problem. You see hives die out, and other colonies move into an empty hive, and then the next year, that colony is dead from mites. Varroa Mites' dominance seems to be absolute if you don't do anything about it." Charlie shakes his head in the affirmative and expounds, "One thing the Bee Informed Partnership has found, and one thing Dennis Van Englesdorp [A famous bee researcher operating out of PennState] said, is that you can be doing everything right, and doing mite treatments, when the fall comes and those around you aren't being responsible, your bees rob the stores of these failing hives that are infected with Varroa Mites and bring the mites back into the colony just after you've cleaned it. It's definitely a different world than before the mite problem." Thinking again of the examples that shed light on the complexity of the bee's situation, I relate, "I was talking with this beekeeper in Port Angeles, Washington and he said that he loses only a third of his hives every year. To me that says something: In that pristine environment, there is something deeper going on, such as poor/unfit genetics."

As we wrap up this afternoon at the facility, Charley paints a vivid picture for us of what a bee-friendly future in the Central Valley (and all agricultural areas) might look like.

In summary, I want to get a feel of Charley's thoughts on current problems and potential solutions. "What are some concerns you see? What is your outlook for bees in the coming decades?" He states, "Varroa Mite is a problem. I'm not sure how we're going to grasp the solution. There is some talk of an RNAi that only hits Varroa Mite, but there's problems with delivery and such. In my mind, all problems are made worse by a lack of bee nutrition." It seems we are on the same page with understanding bee health holistically, and am energized, saying, "I agree. It amplifies all the other problems."

He repeats for emphasis, "They can't fight the good fight if they aren't strong enough. The monoculture farming practice has made honeybee survival a little tough in most areas of the country. It's not just the Central Valley or the Breadbasket, it's the common practice of making sure nothing is growing except that one crop." I understand this, including how the management of much of this country's farmland directly relates to this lack of nutrition. I think about all the suggestions for feeding bees a variety of nectar-and-pollen forage and leaving stores of honey for them to overwinter on, and expound, "I'm aware of a term in agriculture called 'clean fields', where nothing except the cash crop is growing. Many farmers don't even want their microorganisms. Fungicides remove the facilitative fungal micro-root structures which grow on the plant roots that help glean from the soil far more nutrients than would otherwise be possible. There is emerging evidence from Paul Stamets' research that fungicides also deplete bee health because they rely on species of beneficial fungus to keep their hives healthy. They think nature can work in a vacuum, with the web of life all detached."

He sighs and then shares something I hope many engaged in the conventional paradigm will hear with their heart. "Yes. I think that element of control is something I hope people realize is not something we need to do or want to do if we want to keep the system balanced and running. I hope that people in the future realize that we need to allow some sort of room for nature to exist [in agriculture] in order to hedge our bets against current problems. With Varroa Mites, we are surviving. There is another mite species such as Tropilaelaps that has jumped species in the seventies. If we have to deal with Varroa Mites and another kind of mite, we could be facing a real big issue in the beekeeping world."

Charley continues, "I think of planting five percent of cropland in pollinator-friendly stuff. I don't think that's too much to ask. I'd settle for two percent: a space for pollinators to exist, as well as other animals too. That is my dream, I suppose." His vision elevates mine and fills me with inspired energy. I take a few pictures of him for the book and Charley then leaves me to explore the Haven.

As I walk in the Honeybee Haven, I feel a sense of calm and serenity that speaks to the goodness of our partnership with nature. As it is one of the warmer days that week, the honeybees are out and about, visiting many of the still-blooming flowers in the Central Valley's mild winter. As I see them pollinating Lavender, Sage, Cape Balsam, and native California plants, I have an encouraging thought: These honeybees are going to make it. Despite the difficulties with Varroa Mites, monoculture and its arsenal of poisons, and poor/non-adapted bee genetics, I see them, as well as all native bees, continuing to help mankind eat healthfully, even if it requires much adaptation and agricultural upheaval in the coming decades. Their success hinges on much of how farmers manage their land, and upon which kinds of agriculture we reward with our monetary purchases.

I turn back to Dr. Thorp's suggestions on bee gardening. Each of us can have a large impact in our local area by dedicating less of our space to lawn-grass and planting more native plants (for Californians, this also means the garden will be water-friendly). Begin putting good things in your life and in your yard, and the beauty and sweetness will come.

Photography

All pictures are the loving work of the author.
Untitled were taken at the Honeybee Haven in Davis, California
Last page attributed to Audrey Dutton of Food For Rest

Ithaca Waterfall (Page 5)

They Showed Me True Abundance (Page 15)

Sunflower Rendezvous

Niagara Rapids

Apple Culture in Albion, New York

Corn Culture in Medina, New York (Both on Page 16)

Shack By The Sea (Yachats, Oregon)

Brookings, Oregon

Lawyer's Ride in Portland. Pay no mind to the poorly-dressed chap in the shorts! (Page 45)

Grape Culture in Napa, California (Pages 60 & 83)

Skipper and Zinnia

Thistle Celebration

Black Swallowtail on Trumpetweed

Food For Rest – A Christian Permaculture Retreat in Ponca Nebraska (Page 139)

10

Snapshots: Santa Rosa on South

This chapter covers quite a bit of time: From Santa Rosa to Bakersfield, and from thence to Los Angeles and part of the Pacific Coast. You also get a peek into *who* I came out here for, and not just *what*.

Monday, December 12, 2016

After many days of rain at Tim's, the sun had again broken through. He and his family had a long drive to make down to San Francisco, and I went along, as the periods of rain kept causing me to fall behind. I'd often hear, "you sure picked an awful time to bicycle", to which I'd playfully rebut, "I bring the sunny weather with me wherever I go. I have barely had any rain fall on me this entire time." Though this blessing was supernal and unmistakable, the rain was costing me time and so my purist intentions of bicycling the entire length was curtailed in favor of spending the week of Christmas in Bakersfield. I had a bus to catch at the end of the week.

I was dropped off in Santa Rosa, where I quickly made my way towards Napa, California. The vineyards, apple orchards, and rolling hills filled me with delight and I am caught unawares when dusk crept upon me. The only imaginable place to camp was a green blip on the map called Sugarloaf Ridge State Park. I had no idea what it was, but at least it would be far off in the woods. The grade up to the campsite was rough enough to where I'd have to switchback up the road itself to rest and was but a preview of the grade I would tackle the next day.

Tuesday, December 13, 2016

In the early afternoon the next day, after blissfully drifting through a vineyard or two, I ascend over Enchanted Hills, the ridge separating the Sonoma and Napa valleys. Two cyclists out for exercise saw me and my touring setup near the top of the ridge. After catching my breath and talking for a moment, they began indulging in riotous self-denigrating humor, comparing my rugged venture to their "bullshit ride on our ultra-light fancy bicycles". Indeed, the amount of altitude I ascended my bicycle and 30-pound pack over the last few weeks was no small feat, and though I didn't mention it to them, the grade up to the top of this ridge was so steep that I stopped at least half a dozen times. They were a hoot and added to the certain thrill of the route I had chosen downhill. It paralleled the Napa Valley, slowly descending toward the flatlands for nearly ten miles, thus affording an effortless though brisk ride. All the while, myriad small vineyards, flourished with millions of wild mustard flowers, rushed by my sight.

Even during the light rain in Dixon that evening, I got off easy. Perhaps the Three Holy Men of Niagara prayed it so, I have oft mused. I walked leisurely in park until the misty rain cleared (near 11pm) looking for a place to hang hammock or rest all the while. Eventually, I came upon the perfect tree, whose dividing branches leaned more to one side. The distance between the two main branches afforded a 5-ft-high hang which was easy to get in since the hammock now laid almost directly against the trunk. A strange comfort: being held in a tree mid-air to ponder days end as sleep took me. The next day I would travel to Davis, California and conduct some of the most rewarding interviews of this journey.

Thursday, December 15, 2016

Later, there were other opportunities to reconnect with old friends. Take the Jones's, for instance. After fondling away my day in Sacramento, I ended up braving the U.S. Route 50 at night during a blistering rainstorm. Eventually, I was ejected off the highway by a stern but kind policewoman. As we stood there on the highway-side, her in her black uniform and me in my neon-yellow-and-green bicycle jersey, she informed me how common it is for traffic to lose control and slam into the guardrail in a storm like this, and I could have easily gotten hit in that way. 90 minutes later in further rain and getting completely lost, I arrived at the Jones's at 8:30 PM thoroughly soaked and chilled. It took me an hour in front of the fire before I stopped shaking! They were not displeased in the least but were simply happy to have my company.

The next day, the wife showed me her gold collection attained from panning in the creek on their property, along with the miner's lettuce that fed many prospectors in days past. I then helped the husband cut down trees which had "given up" after extensive drought followed by the heavy rains. This extreme dry-wet cycle weakens the root system to the point of collapse or near-collapse for most of his oaks. Time with people like them is one of the sweetest gifts of traveling on the cheap.

Monday, December 19 – Tuesday, December 27, 2016

Marvelous as this all was, it was but preparation for what joy lie ahead. The bus ride to Bakersfield was full of anticipation that gave way to sweet sleep. I awoke somewhere in Merced to the faintest gleams of twilight, a dusty red and purple emerging from the bosom of the Sierras. Newly awake and open-souled, it rendered me objectively happy. I waltzed out in the brisk night during a bus stop, as if life were new, for this valley is where I experienced an entirely new kind of life. Five-and-a-half years prior, I awoke as a Latter-Day Saint missionary in Bakersfield, my task being to spread light and charity to my fellows, wearing a white shirt and tie in the hot Southern California summer.

This time again I came because of destiny, to perhaps begin a new life. Even in the day, a surprising cold greeted me as I departed the bus, filling me with a keen joy mixed with a surreal feeling one might sense upon returning to a land that strangely has one's heart. Southern Bakersfield – Its lined sidewalks, neat grids of sterile lawns, wide roads, and unsavory agricultural and civil pollution – is not the place you'd expect a character of my cloth-cut to find peace. However, every memory from my nine days there rang with the feeling of belonging. I discovered the most peculiar treasure there three years back, though I did not know it at the time: Kirstyn, who became dear to me as the years passed. Not until the month before this trip had I the heart or fortitude to bring her the romantic happiness she longed for.

After a few ravishing dates, and some long walks, many of which were after the rainstorms which offered fresh perspectives on the mountains and life, it was near time to move on. Sadly, these mountains are obscured, so concentrated are the automobile fumes, factory exhaust, and agricultural chemicals of the valley. Though if only for a day or two after these rains of penitence, the fresh views of the southern Sierras may remind those here that there is indeed a better and higher way than what humankind often accepts.

We found ourselves chasing that higher, purer existence our last full day together, up in the high desert of Tehachapi. Tehachapi sits between Bakersfield and Mojave on the southernmost tip of the Sierra Nevada, always receives fresh air, and is one of the few locations in California that enjoys all four seasons.

Our moderate hike took us up part of the Pacific Crest Trail, eventually coming within view of wind turbines sailing along overhead. When we achieved a pleasing viewpoint, we sat underneath one of these giants for a hot meal. Warm, undulating desert floor, orange and red in the afternoon, meets with wintry white and blue endless foothills speckled with more wind turbines. These terminate in the south horizon with the final jagged, sheer stone wall of the Sierras. We were atop a mighty fortress, and upon our battlement, a sacra-dome of safe, open-souled love. Cold gales and warm-breath kisses on cool skin.

Wednesday, December 28, 2016

After many wonderful moments together, my love and I parted for a time. I bicycled the Tehachapi Pass, east of Bakersfield. This epic bicycle ride ends 3700 feet higher than it begins, all in less than twenty-five miles, the last seven ascending the last 1500 feet into the Golden Hills neighborhood. There is a welcome respite from the climbing at Keene, where Cesar Chavez National Monument stands as yet another reminder that ecological and social oppression often cannot be separated and thus must be addressed together. As for the ride, every being and rock in that day was drenched in light and beauty – A lived reality which John Muir called a "terrestrial eternity".

At this ride's commencement, I found bees gathering nectar in a flourishing buzz in this land that knows not winter, giving me a perfect opportunity to do pollination studies. I counted the fertilization rate of seeds in the leguminous pod of a desert shrub species, delicately lit up with flowers painted California-gold. Between Bakersfield and Tehachapi, the bicycle-preferred route offers incredible sights of endless vermillion hills and white-oak savannas, from its beginning to its end. From the western foothills of the Sierras (Edison) to the eastern foothills (Mojave), a distance of less than 50 miles, there is no other place on earth where these four tree dominances stack upon one another:

Willow, White-Oak, Arboreal Spruce / Fir, Joshua Tree. What better, this route follows much of the railroad's path, which is nearly four-times longer from Edison to Tehachapi than the adjacent highway.

Being a "train freak," I absolutely loved the last view of these rolling hills from the Tehachapi Grade lookout. This engineering marvel, completed in the late 1800s, overcame the tremendous challenges of putting a train-track along a pass that far-exceeded the 2% grade maximum for non-cogged locomotive. It was an intense endeavor in both design and dynamite. The Tehachapi Loop is literally a large circle made by the train-track crossing under itself as it winds around a small hill within the pass. The tight curvature, I read, assists the train in slowing down to stop before finishing the descent.

As the sun sank low in the south-west, all the pass is shaded except a burning red portion of the northerly hills. The now-distant hills (from whence I came) beyond Tehachapi Loop still shone and appeared for all the world like a golden garment taken up in the hands of God and unfurled in the mighty wind.

Sunday, January 1, 2017

Under Mojave Desert skies was I come New-Year's. Sleet and terrible wind prevailed that night but gave way to the beautiful calm that reigns over this desert. In the astronomical twilight amid a moist and solid odor of sagebrush – the smell of solitude, I mused – I began the journey to Wrightwood, a ski-town dubbed "the coolest town in California". Incidentally, this ascent to Mountain High Ski Resort at 6700 feet also demanded a 3700 feet uptake from the desert floor, but with a few twists of uncertainty. One was that it had snowed the night before at that altitude and the road could be covered in snow, or worse, ice. Second was that traffic conditions could get grizzly. I both hit patches of ice that occasionally made the back tire free-spin upon pedaling –

a bizarre sensation – and experienced hordes of ski traffic going five miles an hour up the snow-narrowed two-lane road.

After having my fun with the occasional ice for five slow miles, I caught a ride with a group from Los Angeles up to the apogee of the path to Wrightwood. It was all downhill from here, in the glorious cold mountain shine. In the distance, as the cut in the road opened to the great sky, a solid-walled, solitary snow-cloud stood as a yet higher mountain. Though it was but three miles to the Latter-Day Saint church service there, I enjoyed them immensely. I took a stop behind a lovely set of trees and took a snapshot on my phone. Enlivened by the sociality of the hitch-hiking and the expectation of church, I texted my snapshot to many friends and delighted in their response.

What an unexpected sight: a solitary pollinator upon the icy slopes of San Gabriel. It would seem that this creature can bring joy to all climes, and all people. My reception at church as a passerby was the warmest I have yet received in my bicycle touring life.

Everything changed from thence. The ride down from Wrightwood was devastating on a subtle level. It smoldered within me for days thereafter. Not two miles east of Wrightwood, the land was devoured by a September fire and followed me all towards Cajon Junction. In this vast blackened desert, there seemed to be no hope for recovery, no second chance at life.

At this point, leaving behind – though for only a short time – my woman, the uncertainty of her family's reception of me, and the confluence of what all this newness meant for my future, began to reveal its true weight upon my soul. Even gazing at the awe-inspiring white-rock along the U.S. Route 138 and Interstate-15 corridor was clouded over three-fold. My mood, the winter overcast, and the pollution darkened and enshrouded the scene. The traffic added yet more to the destitution of the moment. That oblivious traffic which adds to the blight of mankind – the people exhausted and tortured, yet undeterred in their Sabbath march.

This Sunday afternoon, amid these gorgeous protrusions of raw rounded earth with, as a backdrop, traffic streaming out from decadent Las Vegas to derelict Los Angeles in an endless traffic jam . . . It looked like the end of the world. "These people are just doomed. Totally doomed," thought I. Epic and beautiful, but sad in its finality, such as what one might feel standing witness to a flaming meteor bound for a city caught unawares. That was how the afternoon felt. Perhaps it was the temporary loss of contact with my dear woman. Perhaps it was the emotional let-down that often follows a geographical "let down", a descent from the holy mountain into the world. Whatever the source of the hex upon my mind, that night found me many a time screaming into the cold and the dark. Unfiltered pain.

Monday, January 2, 2017

Or perhaps it was the future asserting itself into the narrow window of the mind that can occasionally foresee and receive it. The despair in these instances can be felt even before the event that precipitates it. Within hours of descending into San Bernadino, the bicycle and all the gear upon it were stolen during a doze in an incredible stroke of fearlessness, or desperation. Though I was less than ten feet away, I never saw the person's face. A strange feeling of helpless rage mixed with relief stirred within me. Something about the bicycle trip started to seem less like a set of wings and more like a cross to bear.

But now, a new burden: walking status. Wretched soreness from countless contacts with cushion-less concrete. Here everything was hard, including people's faces. It felt strange and alien to be in a place such as this. As always, there was human kindness and conversation, but all seemed dulled by the harsh life here. How many people on this planet live in this way? Are the poor in non-industrialized countries all the happier for at least not being told they *are* nothing because they *have* nothing? At least they need not fear one another hour by hour. Regardless of these musings, I found happiness in the fact that all was not lost (wallet, coat, etc.) and my journeys, though cut short, were not entirely over. Events such as this place me deep in thought, and I felt much like Thoreau in this regard:

"[The Australian gold diggings are] the locality to which men furiously rush to probe for their fortunes – uncertain where they shall break ground . . . sometimes digging one hundred and sixty feet before they strike the vein [of gold] or sometimes missing it by a foot – turned into demons and regardless of each other's rights in their thirst for riches. Whole valleys, for thirty miles, suddenly honey-combed by the pits of the miners. . . Having read this, and partly forgotten it, I was thinking accidentally of my own unsatisfactory life, doing as others do; and with that vision of the diggings still before me, I asked myself, why might not I sink a shaft down to the gold within me, and work *that* mine."

- Henry David Thoreau, *Life Without Principle*

After a night in the Gloryland church, which doubled as a homeless shelter, I found recompense at the kind home of Mario. You'll remember him: a fellow rain-crossed voyager whose smiling face greeted me on the bus in Amanda Park, Washington. It was no question upon his return phone call that he wanted me at his place to spend the night and get a meal in Los Angeles. It was almost demanded – such is the typical kindness of those from Hispanic cultures. I was touched, but what I did not expect at all was, upon arriving at his house, him offering one of his four bicycles! We then went out for a brake test. We rode up a steep mountainside path, then down with vicious acceleration, dodging unforgiving trail irregularities. For a first-time ride on the bike, and considering the terrain, I luckily managed to keep my body on the correct side of the handlebars.

Wednesday, January 4, 2017

The day resonated with feelings of spirituality and reunion as I called up old friends. The ride up thru Beverly Glen and down into San Fernando was remarkable. People in California, with its drastic mountain topography and enormous property values, build where few others in America will. Passing hundreds of cars which never passed me later is also a certain thrill.

That night, Mario's wife prepared arepas (potato pancakes) and steak with guacamole for us. I am filled with humble gratitude for his family's kindness. It was truly a balm after the loss that took place in San Bernadino. After this was Camarillo, San Luis Obispo, and Lompoc. All were short stays except San Luis, which I felt held all the perfect hill-and-dale green splendor just for me, as the weather (which I am told refreshed the Northern Sierra snowpack with one immense storm) gave the parched land blessed moisture a few months earlier than normal. This storm also led to massive landslides long the Big Sur portion of the California Route One which took many years to repair.

Friday, January 6 – Wednesday, January 18, 2016

Tim Odenthall, who I knew for many years prior, was as gracious a host in San Luis Obispo, California as I could have ever had. This man has the uncanny ability to create the impression that you are the most interesting person he has met in quite a while. Truly a sincere man, without guile. We visited the mega-dunes of Montana Del Oro in Los Osos my last day. He dropped me off in Morro Bay and allowed me a ravishing 12-mile ride back to San Luis. The seven sisters, formations of upheaved rock from "hot-spot" volcanism, could be seen in all their wild beauty during this ride from Morro Bay to San Luis – Abrupt rock pinnacles arisen amid a gentle valley. Morro Rock is actually the seventh sister of what to me is one of the seven wonders of California.

After a short detour through Lompoc, I hopped a bus to San Francisco. It was wild seeing the Oakland LDS Temple (where this book's cover photo was taken) and riding the county's hydrogen-powered buses. Visiting this city was too quick to be enjoyed but viewing all the homeless souls – souls stuck or unhappy within their sanctuaries of flesh and blood or stuck without a wood-and-glass home – really had me thinking about Christ and the intrinsic value of all human beings. Next was Salt Lake City, Utah, Logan, Utah, and Rexburg, Idaho (to revisit my dear woman), and finally the eternal greyhound ride South Carolina-bound. I am confident in saying that this soul is "traveled out" for a little while.

I couldn't help but think of the below quote, which found me halfway through my travels. Pain, correctly experienced through the lens of humility and gratitude, leads to compassion, openness, and strengthens the non-judging mind.

"Struggle and criticisms are prerequisites for greatness. That is the law of this universe, and no one escapes it. Because pain is life, but you can choose which type. Either the pain on the road to success, or the pain of being haunted with regret."
- Richard Williams, Anthropologist

This epic had doled its fair share. However, suffering is optional. My journeys involved a bicycle accident which broke my laptop and tripod, terrible whiplash, theft of almost all my belongings, being stuck in Vancouver, Canada for a week, beginning stages of hypothermia occasionally, poor sleep (most often at the tail end of the trip), and seeing how wretched humanity can be. All this was pain. Little of it was suffering . . . As I return to South Carolina, my work is not done. I will continue to teach about pollinator decline, win back land to Mother

Nature, and apply to speak to groups from time to time. I also feel prompted to establish garden space for refugees, for pollinator loss is one of many parts of their plight from whence they came, and because above all, habitat loss affects *all species* upon Earth: the only home we will ever know.

That includes us.

11

Sweetness

These next two chapters are not part of the interviews nor are they travel prose. Instead, they highlight the independent research done by the author and encourage the reader to learn more about pollinators and the threats they face. There are also encouragements for those who deeply hunger to change their lifestyle and what they can do to lead, by example, a quiet revolution in their homes and communities.

This chapter is adapted from an article I wrote in September which was published in the *Congaree Chronicle*, a local Sierra Club chapter publication. This was in response to the short-sighted uptick in insecticide usage during the Zika scare of 2016. It provided the initial impetus for the creation of the BeeTheChange.info blog (now discontinued), which became the backbone for the book. There has not been much recent talk about mosquito-borne illnesses, but the buzz on bee populations continues.

"When California was wild, it was one sweet bee-garden through-out its entire length, north and south, and all the way across from the

"

snowy Sierra to the ocean. Wherever a bee might fly within the bounds of this virgin wilderness – throughout every belt and section of climate up to the timber line, bee-flowers bloomed in lavish abundance. The Great Central Plain of California, during the months of March, April, and May, was one smooth, continuous bed of honey-bloom (those of the chaparral belt and lower forests are in full bloom in June, those of the upper and alpine region in July, August, and September), so marvelously rich that, in walking from one end of it to the other, a distance of more than 400 miles, your foot would press about a hundred flowers at every step. Mints, gilias, nemophilas, castilleias, and innumerable compositæ were so crowded together that, had ninety-nine per cent of them been taken away, the plain would still have seemed to any but Californians extravagantly flowery.

The radiant, honey-ful corollas, touching and overlapping, and rising above one another, glowed in the living light like a sunset sky — one sheet of purple and gold, with the bright Sacramento pouring through the midst of it from the north, the San Joaquin from the south, and their many tributaries sweeping in at right angles from the mountains, dividing the plain into sections fringed with trees."

- John Muir, *The Bee Pastures*

On Sunday morning, September 4th, millions of honeybees were found dead in Dorchester County, South Carolina. Naled, a chemical known for its effects on bees, was indiscriminately sprayed from aircraft during daylight hours. The intended target was mosquitoes, but as with most broadcast toxicants, there was collateral damage. This equates to about 80 mature beehives, a relatively small number. However, this is but the beginning of the airborne sprayings, which are intended to continue despite the fact *this chemical retains its toxicity up to twenty hours post-application* (See Pacific Northwest Publication 591: How to Reduce Bee Poisoning from Pesticides). This makes spraying of this chemical totally inappropriate. As with most of our precious things that are forsaken, they are lost from us by degrees.

The people of Dorchester County are taking a stand now. This earth belongs not to a small handful of wealthy individuals, cannot be bought, and cannot be injured without bucking from the forces of life upon her. Degradation of our food-supply is a part of that push-back. Bees are a critical part of our modern food supply. Beekeepers (one including one of the town's fire captains) are demanding financial compensation and there is a general air of contempt among the public. Clifford Scott, a hobby beekeeper in Charleston, expresses his concern: "There were many who lost their entire set of hives, their livelihood, and the county isn't even willing to compensate them for their losses."

An instatement of this careless practice in wetland areas generally would decimate pollinator populations over the better part of the east-coast lowland. Dorchester County is recognized legally as a coastal-zone critical area and deserves better treatment than this. It should be noted: This figure was compiled as it relates to the owned honeybees. There is an unknowable impact on our wild bee populations and other pollinators, some of which are even more sensitive to Naled and other mosquito-control practices. The county would do well cease this toxin campaign, and inform those in the county that they will have to toler-ate the minor inconvenience of mosquitoes until an environmentally-sound solution is found. They might also provide ideas on how citizens can individually mitigate these pests in their outdoor space (citronella and catnip/mint essential oils released into the air can be quite helpful, for example).

I cannot help but recall the prior situations in which bees were caught in the crosshairs during similar foolhardy campaigns against perceived problems. From 1945 to the 70's, DDT was causing major death of non-target creatures such as crayfish, other fish, waterfowl, songbirds, and birds of prey (including our Bald Eagle). One of the targets in this case was also mosquitoes. Impassioned pleas, like those from Rachel Carson and the public spurred by her words in *Silent Spring*, ceased this practice in America by 1972, and populations of these creatures recovered, though DDT's half-life of toxicity is measured in years.

In France 1994, it was discovered that bee populations were declining precipitously. Neonicotinoids, developed by Bayer, were found to be a large contributor. By 2007, research began confirming initial concerns over "Neonics". Research on "Neonics" continued to mount, such Gouslon's in-depth study published in the *Journal of Applied Ecology* in 2013, Chengsheng Lu's two-fold studies in 2012 and 2014, and Matsumoto's *Bulletin of Insectology* research in 2013. When most of the European Union restricted usage of these "Neonics," (including seed coatings like thiamethoxam) in 2013 bee populations stabilized. In Germany, where their ban of Neonicotinoids was more stringent and quickly observed in 2008, bee populations quickly rebounded. However, under proper (though not rare) conditions, these chemicals retain their toxicity in the soil for more than three years (Goulson). Even as early as 1999, France banned Bayer's Imidacloprid-containing Gaucho shortly after its use on sunflower crop, which lead to the devastation of a third of their managed honeybees. When will we value what matters most?

Naled is a mosquito-control chemical that is applied aerially. Pyrethrin is one that is sprayed at ground-level. Both are prevalent, and both unintentionally kill bees and many insects.

There are many examples of people in bee United States showing disdain for indiscriminate toxicant applications in their neighborhoods, homes, and places of recreation. Protests in Miami have shown just how disturbing this aerial spraying is to the public. Reaction by the public in Puerto Rico was so acute that shipments of the chemicals were even shipped back out of the territory (Barreto). And people are right to be concerned. Environmental imbalance tugs at our heart strings because nature is tied into the very fiber of our beings, for we are of her creation. How long will this self-inflicting campaign hold out? Marla Spivak, of the National Beekeepers Association, addressed the decline of pollinators, one aspect of environmental degradation, in this way: "In our collective conscience, in a really primal way, we know we can't afford to lose bees. The bottom line is that bees dying reflects a flowerless landscape and a dysfunctional food system."

Honeybees contribute 22 Billion Euros to the European Union economy and 15 Billion Dollars to the American economy. Nutrient-rich fruit trees such as cherry, apple/pear, almond, blueberry, peach, orange, and tomato (most reliant on bumblebees) depend critically on bee pollinators, as do most fruit trees. Important crops such as strawberry, watermelon, *cucumis* species such as cucumber, as well as all summer / winter squash, honeydew, and cantaloupe cannot reproduce without bees. This pollination need applies to tomorrow's food (seed production) as well as today's food. In addition to these which will become prohibitively expensive, if not lost entirely (7 out of 60 North American crops is one estimate), A whole host of popular foods will have substantial yield-decreases due to drastic honeybee decline.

Squash Bees (*peponapis pruinosa*), a specialist pollinator species for pumpkins, other winter squash, and summer squash deserve special mention. In a study involving observations on 87 North American farms, two-thirds of the yields of squash species would not have occurred without Squash Bees (Cane). These bees collect their pollen and nectar early, relative to honeybees, as the flowers close by early-afternoon. The typical time they collect is in the early morning, often during civil twilight and as early as astronomical twilight (2-3 hours before sunrise) (Cane). This is in direct conflict with the Naled label instructions to only spray at night or sunrise at the very latest. This applies to many other poisons that have toxicities greater than 12-hours post-application, assuming the squash bees are taken into account and the spraying occurs no later than 3 or 4pm.

As American as a painted pumpkin at Halloween is, squash is much more than a nicety. When European settlers first landed in Plymouth in 1621, their winter survival was due in large part to the winter squashes that indigenous peoples had cultivated from wild species. Natives like Chief Samoset and Tisquantum showed them how to grow these calorie-rich staples. Imagine if some of our forebears were more concerned about mosquito-vectored viruses than eating!

Even many staple crops like cotton, alfalfa, sunflower, and canola (rapeseed) are moderately to highly dependent on pollinators. Perhaps we could live off potentially nutritious cash crops such as corn, wheat, soy, and sugar beet. Unfortunately, the great majority of these staple crops have been grown in soil treated with soil-depleting agricultural chemicals for decades, rendering them calorie-rich but nutrient-poor. A surprising number of Americans are not getting their daily intake of nutrients, according to government standards, which lower threshold is based on avoiding illness, not promoting health. Many of those mal-nourished obtain plenty of calories.

The United States has long been identified as calorie-rich but nutrient-poor. There is a direct link between the biodiversity of our farms (the number of species in each acre) and the diversity and nutrient-density of our diets. A lack of locally available and nutrient-rich foods is indeed a risk factor in the susceptibility to disease and viruses such as Zika. People who have little left but staple crops in their diet cannot remain resilient to disease, and as it has been seen, those who succumbed to Zika had frail immune systems.

The catalogue of diseases of malnutrition (such as blindness and beriberi) and their case studies in dozens of foreign regions where Agri-business-powered agriculture (low in biodiversity) has supplanted traditional subsistence agriculture (high in biodiversity) in the local food supply tell of this haunting pattern of human health declining along with biodiversity. Business-as-usual, or even an acceleration of farm industrialization, is not the answer for the plight of these people. If it is not the answer for their nutrition, it is certainly not the answer for us in America.

> Many staple crops do not rely on pollinators, although emerging research is challenging this paradigm, attributing higher yields to pollinator presence*

If an individual wishes to protect his/her family, one can choose to spray their personal property, as well as remove standing water and attract predators of mosquitoes (i.e., via bird houses, nectar feeders). If a business owner would like to honor their customers' wishes, they can choose to fumigate their office perimeter. Before alarmist claims of a national emergency, these were perfectly logical suggestions to most ears.

Regardless of our fears, and the resultant compulsory surrendering of our liberty and health, this generalized spraying flies in the face of individual choice and accountability, as well as our self-evident rights of life, liberty, and pursuit of property. Who will compensate the beekeepers for their loss of property, and in some cases their livelihood? Who can speak to the liberty-lessening effect of ever-increasing food prices – due to illegitimate environmental practices – on a couple who are trying to keep their children healthy on an already-tight budget? And who can place a price on a pollinated field of flowers and the life and well-being of the public?

We have a lot to lose economically as well. In California, where pollinators are on their way out, and where the agriculture business of almonds is propped-up through an intensive and roughly maintained system of mobile pollination services, we see a mostly flowerless land-scape. This stands as a stark contrast to the carnival of color and humming bustle that John Muir so beautifully reveals in our opening excerpt. Colony collapse disorder creates a lack of supply of honeybees, thus increasing the cost of pollination by flat-bed truck delivery of hives. Shortages of bees in the US have increased the cost to farmers by up to 20%. Due to high demand in pollinators, honeybee rental cost has ballooned, and California's almond industry rents approximately 1.6 million honeybee colonies during the spring to pollinate their February almond crop.

These bees cannot stay beyond March because John Muir's flower-calendar has all but faded. After almond-bloom there is no pollen or nectar to feed them, and thus they must be shipped out: a costly and inefficient endeavor. Entire lateral miles of the Central Valley, nearly from the east end to the west end, have become consumed with one or two species of crop trees.

However, agricultural problems can be more challenging than this. In some provinces in China, where pollinators have vanished from the commercial-scale fruit orchards, a strange thing must be done for the apple business. Their loss of pollinators, both wild and cultivated, has reached the extreme end that we could soon see throughout many parts of America if we don't mend our ways. This is due to the same challenges we face: decreasing flower availability and variety due to monoculture and pesticide application, as well as improper planting ratios of male trees to female trees. However, pollination happens, but it is not done by bees at all. Not even by moths or butterflies.

Yes, because bees are totally absent, these apple orchards are polli-nated *by hand*. In a phenomenal study by Partap and Ya, all farmers in six Chinese villages in Sichuan Province use a one-size-fits all regimen of pesticides, herbicides, and fungicides, even though more than half of the farmers know that they kill pollinators. Farmers lose between 12-19

US dollars/person/day due to the need for hand-pollination (where the cost of environmental service of pollination would have been zero). Each visitor to an apple orchard there will tell you there are countless people at work up in the trees (Partap).

My years as a suburbanite produced a typical attitude towards bees: If one meandered into a room, I was sure to be in the opposite corner from it! I feared their sting and their sound put my nerves on razor-edge. Due to this, I managed to avoid a sting until the age of 24. My entire perspective changed once I entered the field of organic farming. I was astounded at the varieties of bugs to be found in the sweet environs of John's Island and Joseph Fields Organic Farm. Among these were our humble bees, including the squash, bumble, and honey species.

It soon became apparent that they had no desire to sting or injure, but rather weren't concerned of my presence at all. It dawned on me, "These creatures aren't a nuisance. They are my coworkers." We were both engaged in the same work: Together we were bringing healthy food to mankind, keeping the society about us in state of peace and sweetness.

A week ago, I found myself bicycling over the first earthen dam in the south, part of South Carolina Highway Six, bordering Lake Murray. I happened upon a sign about bomb testing on Doolittle Island in 1942 in preparation for an attack on the Japanese. I have heard about weapons testing in the Mojave, Cabeza Pieta, and Bonneville deserts, but this historical landmark really hit hard as I gazed out at the hill-turned-island. It reminded me that at the same time the nations of this world began fighting one another, they began also fighting the earth, the soil, even their own food supply.

World War Two brought a host of chemicals that were initially designed to kill man but were post-war rendered to kill other life. Incidentally, bee populations began declining gradually with the shift from nectary cover crops (such as many clover varieties which fix atmospheric nitrogen into the ground) to chemical nitrogen fertilizers in addition to pesticides and herbicides – a continued legacy of violence.

Thus accelerated the unravelling of the web of life – we cannot damage it without damaging ourselves. Airborne spraying of Naled is yet another weave in this terrible tapestry. These losses are gradual, even unperceivable to most, at least until prices of affected food products suddenly become prohibitively expensive. Already, beekeepers are experiencing 2.5 times the losses needed to have serious economic impacts. There is an urgency to make our voices heard in this matter because the airborne sprayings will continue until the love of our ecosystem and healthy food cast out the fear of margin-lying viruses and of nature itself, of which we are inseparably a part.

Doolittle Island has since recovered beautifully from its being bombed. As it stands, it is now a purple martin sanctuary; a place of sweetness, if you will. Many of these flitting birds can be enjoyed upon visiting, and are especially noticeable after a fresh rainstorm, where they can be found dining on one of their favorite foods: Mosquitoes.

We should take extra care of the rich estuaries which have been given us, assuming this solemn charge to be in the hands of each citizen of Dorchester County. If we keep her thus pristine, we can enjoy the benefits of honeybees and all beneficial insects for many generations. We can keep our Earth *and* our society sweet. Must they ever be at odds?

Nine months, 1500 miles, and many jars of honey later, this bee has brought home quite a load of golden pollen, in the form of sage wisdom collected from his travels. I'd like to offer ways on how we can keep the Earth *and* society sweet, for they are intrinsically linked. I share the words of those flowering peoples whose wisdom has blessed my travels.

"The best website to get started [with pollinator protection] is the Xerces Society. It's an insect conservation society that has really gotten into pollinators in recent years. There's a lot of information there. There are books on bee gardening. We just published one recently. It's called *California Bees and Blooms.*"

- Dr. Robbin Thorp, University of California Davis Bee Researcher

"We need to change our eating habits. We need to eat less of the government-subsidized staple foods [such as corn and wheat] and support the kinds of foods you see at farmer's markets."

- Denis, an Upstate New York Farmer

"Consider planting native species in your yards in gardens, as opposed to what just might be pretty. It's always good to give local pollinators native plants. And before you do anything on your property, look into the effects it will have [on the ecology]."

- Alex, Washington Conservation Corps Volunteer

"That element of control is something I hope people realize is not something we need to do or want to do if we want to keep the [eco]system balanced and running. I hope that people in the future realize that we need to allow some sort of room for nature to exist [in agriculture] in order to hedge our bets against current problems."

- Charley Nye, Laidlaw Bee Research Lab Technician

I'd like to finish with some words from my travels in New York. They are mine, but they echo the words of a youth educator I met during my train trip across Canada who opened my mind to this reality: To solve the pressing and novel challenges of this millennium of man, all must play a role in their place of influence. The age of waiting for political leaders to preserve and heal the sacred places where we live and recreate, our sanctuaries in nature, is long over.

"We must teach our children in the deepest ways and empower them so that one day, when they are tempted by economic or cultural forces to sell their birthright – A wild abundance, the environment their Creator has given them as providence – for a mess of pottage, they will have what it takes in their heads and their hearts to protect, so that nature can provide."

*Esquivel et al. 2021 (See Bibliography)

12

An Allegory

"And the harp, and the viol, and the tabret, and the pipe are in their feasts; but they regard not the work of the Lord, neither consider the operation of his hands. Therefore, my people are gone into captivity, because they have no [wisdom]; and their honorable men are famished, and their multitude dried up with thirst."

- *Isaiah 5:12-13 (see also vv 8-10)*

There has been much debate about the harm that insecticides like Neonicotinoids cause invertebrates. I'm talking about eradicated moth populations. I'm talking about bees twitching, dying of nervous-system breakdown, or not being able to find their way back to their hives. And yes, I'm even talking about the decimation wreaked upon the crabs and oysters of our gentle and critical estuaries, from the Chesapeake Bay, south to the Mississippi Delta, and heading north to the Ohio River side of Lake Erie. Many will tell you that the science is not settled, that more research must be done, and that the long-term impacts of Neonicotinoids has not yet been determined sufficient to change government policy . . . as if regulation can only be done by government, and not by conscience.

I have included, as the burden of this chapter, a few links that point to both the homepage of the *Worldwide Integrated Assessment of Systemic Pesticides on Biodiversity and Ecosystems* (WIA) as well as a few selections within. Systemics are genre of pesticides that infiltrate every cell of the plant. This is truly novel: a cutting-edge technology which we may not want to normalize going forward. Following these references, you will find an allegory. This shows the positive feedback loop of environmental disruption and decline, man's reactive behavior, and the false precepts that gave rise to this behavior.

I want to address the idea: the germ of culture and action. Focusing solely on environmental results of the idea may be akin to symptomatic medicine, and thus fail to promote healing. A societal infection has taken hold. It could be named Eco-Manifest Destiny. In essence, corporations such as Bayer, Dow, and Syngenta, along with many of their customers, proclaim, "We shall not rest nor sleep until every wild insect, fungus, and non-monetized plant species is swept from man's domain." The germ is disconnection, or separation: We do not revere and cherish the earth, from whose material we were formed for a wise and glorious purpose.

Home Page:
https://link.springer.com/journal/11356/22/1/page/1

Systemic Insecticides: Trends:
https://link.springer.com/article/10.1007/s11356-014-3470-y

Journal of Applied Ecology:
https://besjournals.onlinelibrary.wiley.com/doi/full/10.1111/1365-2664.12111

WIA Editorial (A quick informative read):
https://link.springer.com/article/10.1007/s11356-014-3220-1

Upon reading these articles, you will likely conclude that if we wish to see and hear again the birds and the bees in all their diversity and splendor, the first step is clear: We as a world-society must cease production, distribution, and purchase of all systemic pesticides, and boycott any companies that sell such products.

Some may argue that as a concerned citizen, citizen scientist – call me what you will – I am not enough involved in the collegiate, political, or business world to boast such confidence or call for such action. One could accuse the author of blazing arrogance, having no understanding of conventional agriculture, the business practices within it, and its impact on our economy.

On the contrary, I assert that *because* I am not tethered to these institutions that I can both see and speak the truth, and fear not what man may do. I have not surrendered my rights of free speech (or more deeply, free thought) in the pledge of allegiance to any corporation's amoral "bottom line". As watermelon specialist and pollinator-lover John L. Coffey says, "One cannot understand a problem with the agri-cultural system if one's salary depends on *not understanding* it." Nobody will fire me, withhold their vote, or manipulate my pension for betray-ing my alliances to their product or service. As for my alliances as an author, I have none other than truth. Though the principle is more important than any examples of it I stand by what is written here. The details within are reflective of reality, though not all details may be found together in every real-life scenario.

Truth is found in nature. Truth is found in the teachings of all the world's religions, whether it be those of indigenous peoples or modern peoples. Truth is found in the lives of good upright men and women who choose the right, both when it would gain them a dollar, and when it would lose them a dollar. Truth in agriculture comes from an abiding trust in the forces that cause it to operate. Essentially, this is solar power, soil structure, soil organisms, and hydrology. This gives rise to properties such as water-retention, nutrient storage, diversity of lifeforms, and ultimately self-renewing fertility. See these in your agricultural system, and you are on the right track.

Observe these processes in an old-growth forest. Harmony, balance, diversity, stability. Life hums in the bosom of Nature, and all the world is one. Water is purified here. A retreat from the heat of the day is always to be found here. Healthy soil, vital to the nourishment of all of Earth's creatures including man, is created here. Sadly, man does not always see its true value. A forestry operation moves in. Instead of selective cutting, allowing a portion of the trees to remain and sustain long-term forest health, a company of wood-hungry mechanical locusts sweeps their broad hand over the whole of the forest, taking away in a few months what it took nature thousands of moon-cycles to establish.

In the current economic mind of the day one best use of this deforested land is a soybean monoculture, for cattle feed in the instance of this tale. After all, alfalfa production declined years ago in this region, and the government will support soy production heartily. One man, who learned from his father, decides to follow in his father's footsteps and purchase this land. "Though there is some debt involved, the returns on soy agriculture are reliable and sound," says the farmer.

There are three problems, however. The second problem is that future rains will wash away most of this denuded land's topsoil. This leads to irrigation and chemical fertilizers as the solution, of which the overwhelming majority will likewise slip away. The third problem is that remaining soil will lose its micro-fauna of bacteria, earthworms, arthropods, and fungi. Monoculture cannot maintain a healthy soil, and often demands chemical fertilizer treatment to continue after a few seasons: a treatment too harsh for our gentle soil organisms. These soil micro-organisms unlock soil minerals, making them available to plants and animals. And what of the first problem?

It is this: Because Eco-Manifest Destiny states that due to our separation from nature, and inherent duty to conquer her, these last two problems will not serve the farmer as a reminder of his place and role on this Earth but will be seen rather as a challenge: an appeal to man's hubris to "up his game". There is no option but to press forward to the crushing terminus of total ecological control. How can a proclaimed worker of the ground, a compassionate feeder of the hungry, believe such a doctrine? How can he believe he is separate from Nature? That he is fundamentally different from the ground from whence he was formed, and act as though his soil were dead, inert, and having no need of compassion?

As the soybean crop next season develops, all is well. Yields are fantastic as pollinators supported by fields untrammeled by heavy equipment and nearby forests not-as-yet logged continue to serve man. Eco-Manifest Destiny prevails, or so it would seem. The plow harvests all the soybeans, a crop of cotton is planted, harvested, and the compacted soil dries out and awaits further cultivation. Winter passes without a concern for the thousands of acres of completely bare ground. Due to the cycles of nature, some years have harder rains in the spring. And it would happen that the time of reckoning has come for man and his fallow field. There has been no preparation – no attempt to restore the balance, at least if temporarily, that has been disrupted by this monoculture . . . and so the rains come.

It rains for days and nights. The finally-concerned farmer thinks it forty days and nights, for he is aware enough to realize that his land is sliding towards the sea. After things dry out on the surface, he begins to plow. The soil beneath is still inundated with water. His heavy tractor tires, as well as the chemical fertilizer deposited behind it, squeeze the remaining air and life out of his soil below the eight-inch depth of his moldboard plow. After a few more passes, the top eight inches is aerated, fluffy, and prepared for seed.

Though lovely on the surface, the farmer does not realize that the water-carrying capacity of his soil has been slashed to less than a tenth of its full potential, due to the repeated compaction of his tractor tires. Nitrogen and Phosphorous have difficulty remaining in soil and on-site without organic matter, and creation of organic matter requires organisms such as earthworms, micro-arthropods, bacteria, and fungi. These dissipated to low levels from the very first application of fertilizer because the fertilizer is too quickly released at too high a concentration, thus "burning" the soil. Only biological forms of fertilizer (Think ammonia or phosphate bound to humus) can be relied upon to maintain soil health.

Because each application of Nitrogen-Phosphorous-Potassium (abbreviated NPK) can only remain in the soil for short periods of time (longer in dry regions, shorter in wet regions) repeated "burning" of beneficial soil organisms occurs many times each year. Even the *Bradyrhizobium Japonicum* that is known to colonize soybean roots, providing free nitrogen to the plant, can no longer be found in the soil. Thus, for good yields the farmer must add even this in the form of seed inoculant. The farmer's soil health continues to decline.

Not long after the spring rains, planting of the first season's soybean crop, and subsequent further packing of the wet weathered soil, the pests move in. Their action is much the same as pathogenic bacteria that prey upon inflamed and necrotic tissue in man. When there is imbalance and sickness, nature has ready those that would return the sick organisms to the soil where their nutrients can fuel new life. The rough treatment of the land has removed a large majority of the predators that keep these pests in check. Plants, though immobile, have intricate methods to protect themselves. They can create compounds that deter pests, and even send out chemical signals to attract pest predators such as the Parasitoid Wasp, Tachinid Fly, or Ladybug. However, plants need a rich soil environment to procure such chemicals. Some processes require priming by soil microorganisms. The soil characteristics needed for this have long been extinguished by erosion and chemical fertilizer applications.

By the end of the second year, the farmer finds himself a little behind financially, as yield income was below farm expenses and the land mortgage payment. He vows to do whatever it takes next year to beat nature at its own game. After taking a tedious exam to receive license to purchase the more tightly-regulated pesticides, he obtains the Neonicotinoids Thiamethoxam and Clothianidin. The first is for seed treating, the second is for foliar application. He dusts his seed in Thiamethoxam and plants, hoping to set it and forget it.

A month later, he returns to his silent field (He has a timed irrigation system that sends out water as needed, according to a weather monitoring program built into the computerized system, supposedly releasing him of the duty to monitor his fields) to find a small population of pests feeding on a small section of his field. He then begins mixing his Clothianidin for spraying next week, which is supposed to be windless and dry.

Next week, the problem is considerably worse, and the farmer sprays liberally, hoping for an improved yield. Two things happen that again disappoint the farmer. First is the reality that the pesticide kills all insects, unlike nature's methods which would have targeted specific insects and let the beneficial ones be. It kills the Bean Leaf Beetle and Dectes Leaf Borer. It kills the Tachinid Fly, its larvae which parasitizes the Bean Leaf beetle, and Ladybug, effective eliminators of many soybean pests. It kills the Honeybee and Bumblebee. His third year's crop yield is already looking to be significantly lower than the first year's yield.

The second misfortune to befall our farmer is pollinator collapse. Both the seed treatment and the foliar spray -- both neonicotinoids in this example -- which toxifies each cell of the plant, have the following effects on both wild and cultivated bees: paralysis and/or disorientation. Thus, there is a loss of pollination for the soybean. Though soybeans can produce well without their pollen being carried from flower to flower, they experience a yield increase as high as thirty percent if pollinators are not chased away by insecticides (Esquivel). His second crop is worse than the first. Destructive trends are continuing around his farm and in his soil with each tree felled on adjacent land and each heavy application of NPK fertilizer.

The final nail in the coffin is the spider mite, a most notorious pest. Ironically, it is *precisely* the Clothianidin he proudly applied early on that caused the spider mite outbreak. The lack of competition brought on by a 'clean field' gives rise to creatures like spider mite first, and its predators later on. With nature's balancing feedback systems silenced by man, in this farmer's current circumstance there can be naught but ruin of the weakened and malnourished soybean crop. If left alone, predators of spider mite, bean leaf beetle, and leaf borer would return and reduce pest damage to a tolerable level. Later applications of pesticide halt this restorative process, however. The third year is his worst yet. "Some years you win, some you lose," says the farmer. He has lost the battle, but not the war. Nature is not his enemy, however. He fights only himself.

Fortunately for this farmer, he is growing a commodity crop deemed essential for production of processed food, cattle feed, and biodiesel. The public's demand for these fruits of soy-production influence more individuals to cultivate larger and larger plots of land for soy agriculture. Public relations campaigns by a wide variety of agricultural chemical companies, as well as land-grant universities which partner with these companies to teach their students to utilize agriculture chemicals, continue to bend public opinion towards more chemical-based and less biology-based solutions. Hence, vast tracts of soybean fields are too productive to fail in the eyes of modern society.

So, by virtue of subsidies of his product, price manipulation of the soy and cattle market, direct federal grants, and consumer demand, he carries on, and for many years his practices change but little. I can't help but wonder about this fictitious individual, whether he is truly fortunate. I think of what Thoreau says of the money-hungry prospector in *Life Without Principle*: "Men Rush to California and Australia as if the true gold were to be found in that direction, but that is to go to the very opposite extreme to where it lies. They go . . . farther and farther away from the true lead and are most unfortunate when they think themselves most successful."

Agricultural upheavals at nature's hand should not be an opportunity to retaliate, but rather seen as a gentle placing of man back *in his place* as steward of this earth, not as its ruler. Systems failures might turn us inward, reflecting on whether we have trust in nature. Consider the might of Mother Earth from beneath and the power from the Sun which she captures, and the power of the life upon her which she supports: the bacteria and archaea, the fungi and nematodes, the herbs and trees, the insects, birds, beasts, and reptiles. Observe without judgment that in each mistake our farmer made in chasing the ideal of Eco-Manifest Destiny, he had a chance to turn back, cease his behavior, and purify his mind of falsehood.

No damage done was permanent. Slowly, weeds would have moved in, healing the ground. Soil minerals would have rebounded through weeds' collection and retention. In fact, what were called accidental 'weeds' yesterday are called farmer-planted 'cover crops' today. Some of these crops even plow the soil along with earthworms and rodents, breaking up compacted soils. Nature's original design is indeed powerful. We can fight it, or we can leverage its power in our favor. Earth can even heal man of pesticides, herbicides, fungicides, and salt (genereally from irrigation and salt-based agricultural chemicals). In the slightly enriched soil environment brought on by the weeds and returning micro fauna, these substances can be bio-degraded into negligible quantities (or in the case of salt, locked away and made insoluble), taking as little as a few months, but sometimes longer than ten years in the case of some Neonicotinoids (Goulson). Pollinators will rebound as flowering populations of wild plants again grace the landscape.

However, this takes time. This takes trust. Do we trust in the biological and chemical processes that have kept mankind well-fed for thousands of years, both in calories and nutrition? Can we release ourselves of the burden of having to always be in charge? Our problems, particularly in agriculture, are self-caused: by our distrust and suspicion, and our feeling of being separated and unworthy. Our broken mindsets: that we are somehow separate from the Earth and walk in front of her, rather than beside her, in this mortal journey.

Let us break from this monoculture of the mind that has caused so much needless suffering. While the physical suffering is apparent, the spiritual suffering may be less so. Let us be like Johnny "Appleseed" Chapman who, as he floated down the Ohio River in a double canoe, created a balanced and stable watercraft by putting as much weight in one canoe (his stock of apple seeds in all their varieties) as the second (himself and his provisions). These two canoes lashed together as one, side-by-side, ensured that nothing on the river would up-end his craft or greatly disturb his experience, come what may. Time is a very long river, and for the uncharted territory before us we cannot negotiate every rapid and plunge it brings with our own ingenuity alone. Thus, let us lash *our* craft to Mother Nature, and travel not ahead of her but as one, side-by-side, so that we can also experience joy and protection in our collective journey (Pollan).

As we are creatures of interbeing, we cannot divorce from the man-ifold splendor of creation without a feeling of immense loss. As John Fire Lame Deer has said, "Being a living part of the earth, we cannot harm any part of her without harming ourselves." A field of flowers is just as much medicine for man's soul as it is food for the bee's body. Huxley's *Brave New World* and Bradbury's *Fahrenheit 451* made some accurate predictions about where man's relationship with nature was headed in the wake of World War Two. Along with them, I plead to let nature back into our lives and, yes, into our agriculture.

Along with Christ, I invite us to rediscover the child within – The child that loved the birds, the trees, the insects, the flowers, and all the Creator's splendor. "Suffer little children, and forbid them not, to come unto me, *for of such is the kingdom of heaven*" (Matthew 19:14). Do you remember that you: the pure being who knew the Earth was good, before it was taught to fear and to put its trust in man's powerful yet reckless arm?

I vow to focus on the cure rather than the sickness. I still exhort everyone to go back and read the links included. I believe we are on the way back up, and that things are getting better, though we be only at the foothills of a great mountain. This has only occurred because concerned citizens have acted, tired of waiting for authorities to step in. Ecological understanding will give power and peace to those who absorb and apply it. We may continue to scrutinize pesticides and test their safety and their effect on non-target organisms, allowing newer, supposedly safer ones to take their place, and hoping for a different result. As the WIA on systemic pesticides reveals, the current affliction from Neonicotinoids on non-target organisms is hauntingly reminis-cent of the days of DDT.

There is a solution that lies far above the realms of chasing bad be-havior. Here is the well-worn path: agriculture-chemical production, application and subsequent damage to mankind and Earth, investigation of said agrichemicals, new legislation to stop the harmful chemical's production, followed by new agrichemical research and production, repeat. These are the ruts in the road we have been tiredly treading since the 1940s, where poisons used to kill man in World War Two were chemically altered with the *intention* of killing target insects, plants, and fungi.

As with political war, so with environmental war: There *will* be collateral damage. The above links give a potent example of this cycle. There is a better way. Trusting nature. Seeking to understand. Uninterrupted time in her endless classroom. Interacting with and helping to heal land, whether under our ownership or that of a consenting associate. Inspiration, and a disintegration of walls of illusion – the illusion of separation. The courage to go forward, acting upon what we have learned.

In the secular speak of the biblical wisdom of Genesis chapter 1, *Homo-Sapien is a keystone species.* Rather than being humans rushing head-long into an unknown abyss of environmental decline, let us be like the Cherokee who lived harmoniously with nature in Southern Appalachia for millennia. Many are still here and carry on these traditions. While they farmed, traded, and from time-to-time warred amongst other tribes like many industrialized peoples, they also were a benefit to the land. One of the big reasons the Blue Ridge Mountains are one of the most biodiverse places on Earth is *because* of the Cherokee, not in spite of them. They brought myriad species from the lowlands like Honey Locust, Chickasaw Plum, Yaupon Holly, Purple Passionflower (*passiflora incarnata*), and Pawpaw and adapted them, over time, to the mountain climate. A greater abundance of insects, birds, and game resulted (Warren).

We are at a critical point, where a severe loss of pollinators could contribute to the reaching of further breaking points in nature, from whence the return will be long and arduous. There will be much hardship in the form of malnutrition / starvation and civil war in the process.

It is said in an article by J.B. MacKinnon that we are living in a ten-percent world, where only a tenth of the life that once was – both in terms of biodiversity and the populations of each species – is with us today. I would say the same goes for us as mankind . . . we could be ten times more wise, more illuminated, more in awe and in service of the sacred than we are at this time. Man and Nature have an eternally-intertwined destiny. Let us begin knowing her, and trusting her. May we be taught by her, emulate her processes in our food production, and serve her. For what we serve ultimately determines what we love, and who we become.

13

An Apple A Day

Not long after the original Bee The Change bicycle tour ended in January 2017, it was time to hit the road again. I drove out to California, got married to Kirstyn, and left out for an extended honeymoon. This created a plentitude of opportunities to meet more earth-lovers. The two following interviews came from this trip.

I'm here on Vashon Island, Washington with Bob Norton (Now passed). He, together with Carol Norton, own Appleheart Farm in Northwest Washington. It was a treat to learn of all he's done and what he enjoys in life.

I ask Bob to gain more understanding of his past, "Where have you lived and what have you researched?" Bob happily relates, "I started out my undergraduate degree at Rutgers, and got a master's at Rutgers. I then went on to Michigan State University and got a PhD there. My first employment in my field was as a professor at Utah State University in Logan, Utah. I was teaching the horticulture courses plus doing research. I did that for about six years, and then an offer came from the University of California in Davis to be an extension specialist. I accepted that offer and was in California for three years, and then an offer came to be a superintendent at the research station in Mt.

Vernon for Washington State University. So, I accepted that job. That was in 1962.

"I spent the next 30 years at Mt. Vernon, Washington doing research and administration of the research station. I then retired in the late '80s. From there we moved to Wenatchee, Washington and spent ten years there. Then, we decided that since our kids are all up and down the Interstate-Five corridor, we would live close by. Carol is my second wife, and her daughter is here on Vashon Island. We bought this property here, built a house and barn, and we've been here ever since." I am delighted by his long and productive life story and reply, "Wonderful. What was some of the critical research that came out of Mt. Vernon for fruit science while you were there?"

Bob provides an insight into his remarkable and multi-faceted career. "I worked on population density in peas, to determine what the best planting density is. There were 10,000 to 20,000 acres of peas grown in the Skagit Valley – and still are but nobody really knew the ideal planting density. I also did some research on raspberry cane control. Raspberries produce new canes each year. If you cut out the old canes, they produce new ones, but sometimes it's helpful if you can inhibit those new canes for a while. We did that with herbicides, and because you want the fruiting canes to be exposed for easy picking before the new canes come up. According to nature, if you don't inhibit them, then you have the new intertwined with the old, which makes picking difficult. I also got involved in lighting research: What is the best schedule for growing plants with supplemental lighting. I went to Alaska and did a lot of research up there on this. I went back and forth to Alaska for a few years.

"As a research manager, I overlooked research on raspberries, strawberries, peas, and at the end of my time there I decided to introduce tree fruits. There were no tree fruits grown commercially in Western Washington at the time. I started bringing apple varieties from everywhere in the world and tested them out to see which were more resistant to disease and insects. I introduced a lot of new varieties that way. I brought the variety Fuji from Japan, and that has become popular. Jonagold came from the east coast in New York. I brought that one out here. I collected apples, both the named and the numbered varieties from all sorts of research programs throughout the United States and tested them out. That's how I got into apples." I respond, "That's pretty prolific work." "Yeah, it was fun. We did a lot of work on growing apples for cider production. There's still a lot of research going on for that at Mt. Vernon."

I am intrigued by cider and cider vinegar production, and I ask, "What's the difference? Is there something else to primarily focus on instead of sugar content?" Bob explains, "Normally you have dessert apples, which are high in sugar content. High in sugar, low in acid. Cider apples are the opposite: high in acid, and not necessarily high in sugar. To make good hard cider, you need the extra acid." I speculate, "It gets the right organisms growing." Bob concurs, "Fungi, yeast particularly. I've done a bit of work on cider production."

> Next, Bob tells me about his research and the agriculture of Skagit County, along with the pollinators that make much of it possible.

Switching gears, I ask, "When did you get interested in pomology, in working with fruit trees? Was it before your university work?" Bob reminisces and replies, "It appealed to me as a kid when my father had fruit trees in the backyard. When I first started at Rutgers I actually started out with ornamental plants, not fruit. The graduate work is what got me into pomology because a professor at Rutgers offered me an assistant position working on the absorption of nitrogen through leaves on peaches and apples and so forth. When I went onto Michigan

State I worked on strawberries. It was mainly on the movement of nutrients through the leaves into the plant.

"I found that calcium, when sprayed on a leaf, stays on the leaf, it isn't translocated into the plant. Whereas phosphorus, within 24 hours, can be found at the root tips." I pick up the pace, and enthusiastically add, "I read some of your thesis; you trace this movement through measuring radioactive signatures, is that right?" "Yes, on radioactive isotopes of Calcium and Phosphorous. We used a Geiger counter. We also used radio-audiograms and X-ray sheets." I later came to think about the translocation of calcium and other important minerals which may land on or be stored in leaves. I realized this movement *does* happen -- more often for deciduous trees, each autumn, and in smaller, more-gradual amounts for evergreen trees. The production of leaf litter under trees is an important part of their self-renewing fertility strategy.

I get to thinking about his previous work and inquire, "When you were in Mt. Vernon in Skagit County, what other plants were commonly grown than those you've mentioned?" "Peas got the biggest acreage. There's a lot of flower bulb production in the United States is in Skagit County. About 75% of all bulbs like Daffodils and Tulips are grown there by a handful of Dutch-descended people. There's a lot of blueberries, which is a relatively new thing. More and more wheat is being grown. Most people think of wheat as being grown in Eastern Washington, but different kinds are grown here." "With the exception of the wheat, it sounds like Skagit county horticulture greatly relies upon pollinators." "Peas don't seem to require pollination. Fruit trees do, and others, such as hybrid seed crops. Some are done in a closed environment where they introduce the pollinators. Pollination is im-portant everywhere, but there are not beekeepers all over."

Steering the conversation back to pollinators, I call upon his long-term residence in the area. "In the 30 years you were there in Skagit County, did you see an increase in the demand for hives for pollination?" "I don't know, but we had plenty of pollination for our fruit trees, although we would bring in a few hives. You need about a hive per acre. You have your ups and downs. You have colony collapse and other things."

Thinking about colony collapse, I add, "They don't even make it back to the hive. There's no way to biopsy because there are no bodies to be found. There has been some research done by scientists like Matsumoto in Japan. He looked at homing times for honeybees at differing doses of insecticides. It didn't cause a lot of bees to go missing but he did notice that a very small sub-lethal dose of insecticides in an area would cause them to return to their hives at a slower rate." Bob adds, "People blame pesticides, but nobody has been able to pin the loss of bees to pesticides alone." I adjust, "Well, not alone. It's always a lot of different factors." "Right," Bob concurs.

I provide an example of the complexity surrounding bee health: "Expert mycologist Paul Stamets gave a talk on honeybees in 2016 at a conference in Vancouver. He's a graduate of Evergreen College in Olympia and has done lots of research in this state's biomes. In this conference he shared his discovery that most honeybee colonies, just as a rule of their existence, rely upon a certain fungus. It's something that nobody sees unless you're looking for it.

"He found that in the absence of that fungus, bees were far more susceptible to all these other pressures that get more attention for bee mortality and their seeming need for antibiotics -- a common practice in commercial beekeeping. I'm not convinced that's a proper solution. However, more often than not, when in his research they restored that strain of fungi to the hives, the bees' resilience greatly improved (Stamets). It's so interesting to hear that because fungus is usually seen as a bane that farmers want to get rid of. It's not thought of something beneficial, which is a shame." Bob summarizes, "Some kinds you want gone, some kinds you need to have."

As we get to thinking about bees, Bob shares how he uses pollinators to maximize the yield in his fruit orchards.

I am intrigued by his mason bees from the moment we arrived here and wish to learn more about them. "So, I noticed you have some mason bees here and some honeybees. When did you acquire those?" "I've had the mason bees for at least ten years and at first, I was getting them from a mason bee supplier, but now I have plenty of my own, stored in what I call 'hotels'. I have about two dozen hotels now. My only problem now is having to find empty hotels for them. When you put out the full mason bee blocks you need empty ones for them to lay their eggs."

I inquire further, "So, what is the life cycle like for the mason bees?" Bob explains, "About three months, out of the block. The bees come out in March. They go and get pollen, fly back to an empty block, crawl in the hole, deposit an egg, and build a wall, and go and repeat the process. About five or six eggs are deposited into one of those chambers. When they fill them up, they seal off the entrance and they die. The males come out first, and they wait for the females to come out. The females are a little further back in the chamber. They fertilize the female, and then they die." I humorously conclude, "It's a hard and fast life for the male mason bee." Bob chuckles and confirms, "Definitely a short life. And when the female bee lays her eggs, she dies." I echo the previous sentiment, "An ephemeral bee species." I have read that mason bees are far more efficient pollinators in that their diet is much higher in pollen. Though they operate in a rather small window of time, their usefulness to fruit/vegetable production cannot be understated.

I then ask about his honeybees. "I just have one. It's owned by a beekeeper. She takes care of the hive. She only looks at it once or twice a year." I seek for indications of their health, "Does she come to apply the miticide every year?" Bob matter-of-factly states, "She uses no chemicals whatsoever. She doesn't treat for Varroa Mites." Surprised, I ask, "Really? And about how long has that been going?" "Three years." I am pleased at the concept of domesticated bees lacking the fragility which I see in many industrial managed hives. I relate, "My understanding is that in areas dominated by corporate agriculture if you don't control *Varroa Destructor* your hive has two years tops, and that's if you're lucky. Then again, this is not that kind of area." Referring to the equally resilient mason bees, Bob offers, "I've seen them on the back of the mason bees, and they'll get on the developing larvae, on the block. They sometimes emerge with the Varroa on their back, but the mason bees seem to get rid of them."

The idea of Varroa as an insurmountable obstacle again dismantles in my mind and am reminded of what Austrian permaculturist Sepp Holzer says regarding his bees. He places his honeybee hives amid tall plantings of mints and other rich-scented herbs. Holzer relates in *Desert or Paradise* that the phyto-chemicals and resins (collectively called propolis) gathered from the nearby herbs are used by the bees to either "treat" their hives against Varroa Mites or create an inhospitable environment so that they never get a foothold on the colony to begin with.

After being amazed by his statements about Varroa resistance / immunity, I think about the variety of apples these pollinators are working with Bob to create. "I understand you're seeking a patent on a kind of apple you're cultivating." Bob beams, "Yes." "Please, walk us through the steps of creating a new apple variety."

Bob shows us how he went about making his own apple variety, and how you can too.

Bob enthusiastically explains, "Okay, well the first step is to decide what the parents are going to be. Say you want an apple that's like a

Honeycrisp. Well, you might use Honeycrisp as one of the parents, but maybe you want more disease or insect resistance or a certain-sized apple, so you pick the parent that has that characteristic. Then the cross can be made either way: The pollen can come from either tree, and then that pollen is deposited on the stigma of the other tree's flower. You self-pollinate the flower. You save the fruit, and when it ripens, you take out the seeds, let them mature, and go through their rest period.

"In January/February you germinate that seed and start out with seedlings. Now if you let that seedling go on its own it will take six or eight years to produce fruit. So, oftentimes you'll take buds from that young tree, put it on a dwarfing rootstock, and you'll get fruit a little quicker: one to two years sooner. Then you grow a thousand of them. One in 10,000 might be the one you want." I offer, "But even then, every time you get new apple seed you have to shuffle through so many genes to get what you want. I mean, apples have such a large amount of genetic material." Bob confirms, "Oh, right. Apples are very heterozygous. That's why you need so many samples, to get one in ten thousand that's better than its parents."

I observe that he has been working on this a long time. "Yep. I've only had a hundred, but I selected that one that came closest. It's got pretty good qualities, but it probably will never become an important variety. It could be a backyard variety that nurseries here can sell." I add, "As a novelty because it was developed here. That's a plus for sure because it's used to this climate." Bob continues, "Good eating quality, good storage quality, but not as big or attractive as a Red Delicious." I tell him that I've picked this variety and that I think they're moderately attractive. I tell how I appreciate that they're not uniform red orbs. He explains, "You get more variation with this apple. Some have a strong blush, others don't. That is not good from a commercial standpoint. They want every fruit to look the same."

I ask, "How do you feel about that: the homogenization of consumer tastes?" Bob explains, "Well, it is a commercial thing. That's why we have farmer's markets so farmers can produce, on a smaller scale, varieties that have some unique qualities that people would like. But this apple will perhaps be a variety that backyard growers can grow, and there's a lot of those people here in Vashon. In the fruit club, we have a couple hundred, and that's just on Vashon Island. In the state of Washington, it would be ten thousand." I encouragingly add, "Everybody doing a little bit. That's all it takes." "Yes." "And I think it's really important to have some varieties that have relevance in a local area that is used to those circumstances." Bob bounces off this idea, "That's unique for this particular area . . . and that's what this one is like."

I postulate and summarize, "Local resilience in food production is critical for local pollinators and other wildlife. So, it takes three different parents – a mother, a father, and a nanny – to get the kind of tree you want." He affirms, "The dwarfing rootstock, to get precocity. It's not really a parent, but you still need it." As an aside, I tell Bob, "That's my new favorite word." "Precocity?" "Yeah. Unless you're a scientist, you wouldn't know that word comes from precocious, which is great." Bob laughs at this and uses it in a sentence, "You hope your children will be gifted and smarter than the other kids in the class. You hope they'll show precocity."

We wrap up by talking about legacy and hopes for the future

In closing, I ask, "What's something you'd pass on to the next generation of farmers? If you could pass on what you learned?" Bob takes a moment, "I've thought about that over the past years. I don't know if there's anything unique that I'd pass on. My family is not involved in fruit growing, so it be wouldn't through my family, but I might have a little bit of influence with the people on Vashon Island, so they might continue to grow. Maybe they'll use my variety of apple.

I offer, "Perhaps they can continue to perfect your variety and get it more homed in to find those desired characteristics." "That might be something that could carry on. I'm not really expecting a legacy." What a humble man. I certainly see a lot of his work maintaining its importance for decades to come. "Well, I think of being the means, or part of the means, by which some of the aforementioned apple varieties are commonly grown in this part of Washington is a legacy. That's neat. I didn't know that before talking with you. I think that's pretty important." Bob deflects, "We'll see." I laugh with him, accepting his modesty, "Yeah, we'll see. Thanks so much for your time." Bob ties up the conversation by saying, "You're welcome. I've enjoyed talking with you."

This has been such an informative and inspiring conversation. While me and Bob didn't cover any things that we can do to improve the pollinator situation, I believe the understanding of how to create local varieties of fruit trees, along with the ability to both raise and host multiple kinds of pollinator species, is a critical aspect of the nature-healing these pollinators need. Though Mr. Norton has worked for decades within the agriculture universities which prize homogenization and hyper-specialization (think of entire valleys dedicated to the growing of one species, spread out over dozens of counties, each valley with its main dominance of one or two species. This is an accurate picture of the source of much grocery-store produce) his ability and passion for creating local resilience is inspiring.

Earlier in the week, I and Kirstyn visited the fruit club (they meet at a different person's property each week), wherein dozens of individuals on the island were saving seeds and developing varieties specifically suited to Vashon Island. Bees, both wild and domesticated, depend on this kind of work, since the resilience in locally adapted plant varieties buffers against pest and disease pressures that can wipe out the hyper-specialized commercial varieties with greater ease. Creating local varieties / landraces is the crucial way pollinators affect their neighboring flora for their own benefit. Let us partner with them in this creation of local resilience.

It is said that the planting of a few dozen trees was a common act early American settlers and was used to announce, "We are here to stay". I believe that the Earth desires this long-term commitment from us in our valley, on our acre, or even our 3,000 square-foot back or front yard. Wherever we find ourselves, we can show forth a long-term concern for our environs. Bob Norton has exhibited a fine example for us to emulate.

14

The Beekeeper: Giver of Sweetness

"A lot of people say, 'my vote doesn't matter'. Whatever it is you believe, you can express that as you spend your money. I think when we're all kids, we want to save the world. Farming allows you to see, 'Okay, maybe I can't save the world, but I can do a good job on these three acres.'"

- Kathleen Shannon, Beekeeper (2017)

While talking with Bob Norton, we learned about Mason Bees. We now have the opportunity to learn more about honeybees. I'm here in Snohomish, Washington with Kathleen Shannon, an energetic beekeeper. She has just finished her first year caring for bees in coastal Maine. We got to talking deeply about some poignant topics, and while parts of this conversation are an affront to the economic status-quo, I have chosen to publish the entire interview in full. All economies are but a sub-system within the sphere of the natural environment, and cannot function without the resources/services, such as pollination, she so graciously provides.

The word 'political' as it is used here refers to making change within one's community, rather than its common connotation of affecting government policy on the state or national level. Or, as David Holmgren put so elegantly: "Those who act [successfully in activism] do something that benefits themselves first. Any movement that's going to bring about change . . . is going to be a benefit to those who act first . . . There are a lot of disincentives, but there's also some incredible advantages in being early adopters." Each one of us, as people in the affluent world, have enormous leverage through what we choose to financially support. He continues, "I have found that apparently by just living a simple life, that [self-reliant] model is more inspirational and provides more leverage of influence than doing anything else." (Arigona)

I begin, "Kathleen, what first got you into this practice of bee-keeping?" Kathleen explains, "I have been working on organic farms for three years prior to this. A lot of that came from a social justice perspective. The pollination aspect of beekeeping is sort of aligned with the justice issues. It feels like a political act to keep bees." I am intrigued by her explanation. I think about the often top-down connotation of the word 'politics' and the many bottom-up concepts utilized by those who espouse 'earth stewardship'. "Well, it's a question about the commons, which is very political right now; the idea that there are things that are owned by commonwealth, which everyone has equal access to. Now so much of that common land (we have our State and National Parks, Bureau of Land Management land, etc.) seems to be for sale to corporate interests. It's not as much reserved for the commonwealth of the people who rely on those lands for environmental services." Kathleen responds by explaining that beekeeping doesn't see human property lines and the idea of maintaining the health of the commons as integral to bee health.

I affirm and she continues, "However, it seems like a political act to keep bees: I mean, to make my own sweetener, all while doing something that is good for my community. It frees one greatly from the food system." I think about the implications of what she has said about producing one's own sweetener. There are indeed political echelons and armies of legal defendants employed at the protection of the corn, beet, and cane-sugar industries. To circumvent this system through gathering sweetness from the nectary bounty of the commons is truly, in its own simple way, radical. I add, "All that is done on no land, if you think about it. There are cities in 2nd and 3rd-world countries which supply most of their honey in-city, with bees harvesting only from the nectar of flowers planted by people and flowering weeds. There are no fields for beekeepers to place a hive, just a very small backyard."

Re-iterating the relationship between beekeeping and the commons, she adds, "There can be lots of trees in parks which feed bees as well. I really feel like beekeeping strengthens your community beyond just your vegetable garden and your backyard." I agree and explain for the purposes of the reader, "Bees forage generally for two miles in any direction from the hive, but in the spring when they are in greater need for nutrients, they will fly up to four miles to get a variety of species in their diet. They know to eat more than just corn flakes, as it were." Kathleen excitedly adds, "Also, it's a localized product. Eating local honey can help with allergies because the bees are gathering pollen from your immediate area. It's such a good representation of your bioregion." I summarize, "Yes: It's a strong expression of place." "Well put," she says.

> Kathleen shares with us a look at the abundance honeybees
> enjoy in Coastal Maine

Interested in her bees, I ask, "Could you walk us through the season for bees in Maine from the first nectar flow to the fall?" She relates, "I got my bees April 20th. Maine is pretty cold. We were close enough to the ocean, so it wasn't as cold, as the ocean stabilizes temperature extremes. It's a quick season, and the winter is hard on bees because it is humid and cold. The first flow is red maple, which is not before April. Dandelion is the next thing to flower. Goldenrod and Aster follow, and Buckwheat is one of the last major food sources for the bees. The last time you should harvest is Labor Day because you want to leave them with enough honey for themselves. I had them closed up for winter by Halloween. They'll slow a lot by then, but they still get around." I ask for clarification by way of assumption, "I would imagine they aren't active every day before you close them up. There are warmer days where they forage and many strings of days where they don't." She affirms and says, "They slow down their foraging, but on the high season it is enjoyable to watch them buzzing about from flower to flower."

.

Changing gears, I think of her ongoing education and ask, "What were some of the stories you looked at for your 2012 paper?" She replies, "I was writing about colony collapse disorder. At that time the studies weren't as conclusive. They would each show different conclusions." I surmise, "Particularly because that's what they were looking for. Usually if a scientist with a hyper-honed-in degree is looking for a specific kind of problem, they're predisposed to finding that one thing. All these scientists are likely finding something correct, but just not looking at other data." Kathleen continues, "It was a systems class I wrote this paper for. I was trying to determine why this system was collapsing, so of course there's so much that ties into it. Monoculture, trucking bees for thousands of miles, pesticides, and more play into the problem." I agree, "All of those deserve more attention."

Returning back to her bees, Kathleen says, "I know beekeeping is helpful for my community. I got my supplies after my tax return. The beekeeping community is really tight, and I like what that taught me about being part of a community." Thinking about how honeybees are not as resilient as wild bees, I opine, "I think they're the canary in the coal mine. If there's an agricultural problem, honeybee mortality will usually be the first to tell you. There have been multiple instances when a new agricultural chemical was released and half of the beekeepers in the area raised hell because their entire hives were decimated. It's awful to find out something doesn't work well because of honeybees dying, but I think it's good to have an early warning system built into our agriculture." She expounds, "It's kind of a scary thing, but we can predict our environmental health by looking at their health. They're going out in the natural world to places that you wouldn't suspect or look for. They bring it all back to their hive."

Building on this theme of interconnectedness and variation in the natural world, I add, "I've done a good bit of reading on what's called 'Best Management Practices' for large-scale growers. For instance, the owners of vast tracts of almond orchards are given advice such as, 'Make sure there aren't flowering weeds within the breadth of your hired hives'. So, a lot of these farmers are living within a pipedream: the idea that one can control every life-form within one's region; and that as long as there are zero flowering weeds, the agricultural poisons won't land on them, and the bee-keeper's bees won't land on them, and the bees won't get poisoned. So, like you said, the world of beekeeping helps you see things in their inter-connected reality. If all you see is money, then you could follow these management practices and think that you're fine. But if you are doing it for the love and passion for living things, you see that it does not work for all the beings involved."

> Kathleen has discovered her pathway of peace, and it does
> *not* involve saving the world

Kathleen concludes, "I was saying before that I believe it's a political action to keep bees on a small scale. I believe in voting with my dollars. I try to never shop at Walmart or McDonalds." I reflect, "You shop where your values align with your purchases, to the best of your knowledge." "Right. I try to be really careful with that stuff. A lot of people say, 'my vote doesn't matter.' Whatever it is you believe, you can express that as you spend your money. I think when we're all kids, we want to save the world. Farming allows you to see, 'Okay, maybe I can't save the world, but I can do a good job on these three acres.' The cool thing about beekeeping is that I can make a palpable change within these five miles with these bees, and I can make a difference in my community. That feels right and feels worth doing."

In my mind, Kathleen has just taught us one of the most potent lessons of this book: Each dollar spent is a vote. In fact, I would propose that this has a more powerful and far-reaching effect than what we decide at the ballots every year or so. This concept was touched upon in the *New York* chapter and reiterated in *Sweetness.* If we wish to inhabit a more biodiverse world, we would be wise to financially reward those who grow and tend a variety of plant and animal life, while at the same time removing our financial support from agricultural enterprises that vacate the world of community, beauty, and health.

The biodiversity of our farm ecosystems is a direct reflection of the biodiversity of our diet. When we support our local family-farm producers first, we ensure this positive reality in our own neighboring regions. A critical mass of concerned citizens doing this in their region makes a career in biodiverse local-based agriculture highly attractive and may provide the financial impetus for many more of our neighbors to engage in a food-production business. This is a powerful act of community autonomy. This is something Kathleen understands on a deep level as she bolsters her community by keeping bees. It is just as we discovered in the interview before: Everybody doing a little bit can create a powerful wave of positive change.

15

The Experiment of Economic Decline

"I feel like this is what the Millennium will be like: Where the air is clean, and the streets aren't clogged with traffic and noise" says my then-wife Kirstyn. In the Christian tradition, this Millennium is the 1,000-year period of Christ's presence on the Earth where war – supposedly man-to-man as well as man-to-nature – will cease. Tikkun Olam, or "The healing of the world" in the Jewish tradition, is to take place during this same time period of divine reign.

We both reveled in the die-down of busy-ness during April, May, and June 2020 while the experiment had first gripped the American West. Indeed, for the first time in our years in Salt Lake City the mountains were clearly visible for months at a time, their jagged topography rendered in all their gristly detail. There was little to no smog to damper the rock and snow songs of the Wasatch Range, one of the newest mountain ranges in the Western Hemisphere. We enjoyed the grandeur of those mile-prominent stones for weeks during our walks and bicycle rides in the city. It truly was a blessed time for many people as well as many pollinators.

As written earlier in this book in "An Allegory," when we take off the pressure, life re-emerges and can again offer its gifts. As many monoculture fields laid fallow from the slump in economic demand, flowering weeds returned and fed the bees and butterflies. This in turn feeds birds, rodents, and reptiles. While labor-supply and demand for extractive industries faltered, the Earth's natural resources in wood, cleanliness of rivers and streams, and soil fertility rebounded. As county and state governments sent roadside-maintenance employees home, the brush in many regions has flowered along roadsides, growing further into maturity than normal and feeding pollinators where in other years they would have been slashed down before their time.

David Holmgren – co-founder of permaculture, a regenerative land practice – has said in his essay *Crash on Demand* that it is *only* an economic downturn that has been concretely proven to reduce ecosystem decline as well as green-house gas emissions. One should not downplay the value of renewable energy, energy-efficient city planning, or any other "green" innovation. It is simply that of all their many concretely-proven benefits – financial, resilience of national or local economy, quality of life – reducing greenhouse gas emissions and damage to nature aren't among them (Arigona).

I am not downplaying the tragic emotional and even spiritual consequences for humanity as people have been torn from places and patterns of belonging and purpose such as social gatherings and employment. I have been quite vocal about these aspects in other writings. It is however a forceful manifestation of the truth that 'as one door closes, another opens.' In this journey of life, opportunity cannot be totally removed from us, but only shifts from one form to the next.

For the remainder of the book, I will refer to the months of March 2020 to August 2021 as 'the experiment'. I am just aware as others – perhaps more so – that much else has been at play. However, I propose that the generally observed results of economic contraction on rebounding ecosystems may well be the greatest, as well as unexpected, discovery of this time period. Or, at least for the purposes of this book, it was the greatest take-away for the author.

A collective awareness is rising among us that nature desires our deeper attention and spiritual energy. Many who maintain places of outdoor recreation have either felt this energy or recognized the upsurge in desire from the public. Our next interview covers the experience of one State Park manager who has noticed this experiment at work.

16

The Ten-Thousand Hands

"Every permaculture landscape looks different. Ours is specified for heal-ing people and giving them rest. We teach people that come to our place how we do what we do and how nature and our Creator has guided us to live in a state of abundance, both spiritually and physically."

- John Dutton (2021)

This interview was conducted online with John and Audrey Dutton, of Food For Rest, a Christian-based permaculture retreat in Ponca, Nebraska. It was indeed a rare treat to spend time on their 80 rolling acres last summer with my family and theirs. Getting to know the Dutton's has been remarkable and uplifting.

As this past year-and-a-half has shown (at the time of writing, March 2020 to August 2021), worldwide societal changes touched al-most every facet of the planet. Ecology and our relationship to it are no exceptions and are in fact some of the most memorable changes encountered during this experiment. For metropolitan peoples, the stark increase in visibility and breathability of the air was perhaps the most notable difference. Many locations the world over saw dramatic reductions in pollution of all categories, be it industrial, agricultural, or commercial. As pressure eased on adjoining natural areas, both aquatic and land-dwelling species saw an increase in their populations and breadth of distribution.

The focus of our talk today is along this vein. John and Audrey have had lots of textbook and real-life experience with permaculture and use their land to teach the many visitors that pass through each year. John Dutton is also head groundskeeper at Ponca State Park and he will be sharing many insights from his time tending nature there.

I am interested in their keen observations of the land around them. Using that as a starting point, I ask, "So, Dutton's, what are your perspectives how this recent time has facilitated ecosystem recovery and stoked people's interest in nature." John begins, "People definitely kept off the streets and in seclusion which is detrimental to social connectivity but offered more connections in a different way. The ecosystem is designed to repair itself. When it is not being suppressed or cut down or blocked off, nature will come back in a beautiful way. Even at nuclear bomb sites and industrial areas, nature has found its way back into the landscape. It is the only system that could bear such oppressive restraint and still survive."

Audrey offers a complementary insight, "In our region, of the eleven years that I have lived here (two miles south of Nebraska Highway 12), last winter was the first time that I have ever seen deer on the south side of the highway. Pheasants, turkey, and quail have multiplied, and I feel that's a result of the lack of traffic on the highways." It is wonderful to hear this and I say, "I think you both are an example of 'iron sharpens iron' in terms of permaculture and natural history.

"Audrey, it has long been known that busy highways sector wildlife into smaller corridors, where the likelihood of mating opportunities and resource procurement severely declines. Also, traffic kills and most field biologists realize it. The highway dropping off in vehicle 'population' would allow species from the Missouri river (north) side of the highway to expand their habitat and grow their populations through their now-expanded land-base."

John Dutton is the groundskeeper manager at Ponca State Park. He shares some of his rich experiences there

I turn to John and ask, "What are some of your observations/ experiences from Ponca State Park?" Ever ready to share, he remarks, "I noticed that people were coming in for programs and work asking about the different biomes that the park has like Wildflower Prairie, Forest, and Oak Savannah. I've had discussions with the head naturalist who makes up programs and provides material for our visitors' learning experience. I've been speaking to her about the importance of soil health and how we can contribute to that in the way we work. I had suggestions about making seed bombs which are mud balls with different seeds in them that you throw into places to have greater diversity come up. And I expanded our nutrient recycling by using horse manure and turning that into compost by mixing it with wood chips and then also implementing small compost turning bins at our pollinator garden."

I infer that the ecological interest of the park visitors has been increasing as well. "I'd imagine with your seed bombs and pollinator garden efforts you've seen increases in pollinator diversity." John affirms, "As we have planted more wildflower strips at the park, I do see greater diversity in the types of insects and animals that come through which sparks the curiosity of people that walk by."

I ask, "Any moments with patrons that were memorable?" John relates, "Today, the naturalist stopped me and had me play the role of educator for about an hour for a grandfather and his grandson trying to earn a badge in conservation. I took the boy and his grandfather out to our backwaters, which is almost like an island in the river and supports a wide variety of wildlife because it has prairie and woodland all mixed in there. I showed him how to do a soil test by mixing soil and water then gave that to him and explained different soil types and how we can design places for conservation and the objective of those conservation efforts."

I warmly smile, "That sounds like a special experience. Your work must be so gratifying because you are helping to bring people closer to being a nature healer every day. All this work will certainly give nature a large push towards recovery within the park. About when did the foot and car traffic pick back up for the park?" John explains, "The foot / car traffic didn't really increase. The interaction with nature simply increased because people couldn't go into the indoor exhibits. So, they were spaced-out outside and were getting more time in the actual nature of the park. As for some additional nature-healing efforts, at Ponca State Park we're trying to conserve piping plovers by creating more habitat. In so doing, we created more habitat for beavers, cranes, eagles, and more."

I ponder and inquire, "Do you think that an upsurge of naturalists in the next decade could be a happy consequence of the changes in the last 18 months?" John reflects on his perspectives gained through permaculture and reflects, "It's hard to say. There are those who know a part of the natural system and just see it as a part. They often compartmentalize that part in their minds, but as a permaculturist I like to take a step back and look at things in a holistic way. If we had more of that kind of thinking it would be a great benefit to the world."

I expound and consider, "I really believe that as one spends time in nature, celebrating diversity becomes more of a natural inclination. We live in a very standardized modern world, and so I think the increased exposure people are getting, combined with many ecosystems currently rebounding and having more of their 'glory to display,' could make for some real growth: of both the ecosystem, and of human compassion and awareness for it. And I do see that we need a lot more generalists who can positively interact with many families of beings (trees, insects, birds, amphibians, etc.), not only more specialists." John inserts a thought, "Generalists . . . I like that. I think most farmers are generalists."

Continuing my train of thought, I make a medical metaphor. "My father has half a dozen doctors he talks to, and they are, at best, meagerly informed of each other's discipline and what work they're doing with my father. Whereas the only person I see is a chiropractor whom I've known for many years. He knows so much about myriad systems of the body, and how they work together. He also does physical-therapy-type work on me while he's doing a chiropractic session.

"I feel he can get so much more done because of his 'cross-training'. Relating it to farm work, if you only know how to work with the crop, you'll be far less equipped to produce a nutritious and healthy harvest than the farmer who knows how to work with the crop, the beneficial insects, the soil microbiota, farm animals, etc." John affirms, and I explain that we don't have to be experts in one field, we just have to think holistically as John Dutton as said, and we can accomplish much to help improve the natural world by obtaining a practical knowledge of working with many different kinds of life. I conclude, "It doesn't take a master's degree in entomology, hydrology, forestry, and animal science to produce a beautiful landscape that meets the needs of people and our non-human neighbors."

The Dutton's share with us their efforts to create sacred space on their 80 acres

John agrees and shares the special opportunity he has as a permaculture practitioner who has access to 80 acres in the heartland of

America. "Speaking of farmers, we are in a unique situation. We are developing our own little system of intense biodiversity amid a monoculture-pressured landscape, dominated by genetic and mental / methodological sameness – or in other words, conventional farming."

As we are both believers in Christ and His gospel, I can't help but think of the spiritual comparison of how we can maintain a peaceable walk with God in our family or home space while amongst a modern culture than is increasingly irreverent of the sacred. I hopefully consider, "I'm sure you've been empowered upon realizing that our own little changes in our own space can quickly create resilience to the outside pollution and myopic behavior. How does this all fit your mission at Food For Rest?"

John and Audrey then beautifully relate their vision for Food For Rest. "Every permaculture landscape looks different. Ours is specified for healing people and giving them rest. Giving them actual physical food as well as food for thought: We teach people that come to our place how we do what we do and how nature and our Creator has guided us to live in a state of abundance, both spiritually and physically. We plan on expanding our Airbnb short-stay platform. It has been a good income source and a good way to teach what we hold so dear to us. We want people to come and experience that as they heal the land the land heals them."

I am inspired and recall, "I remember feeling like life could be better after a few weeks' stay with you last year. It was a beautiful and transforming experience. I think the value of any great spiritual work could be measured in this way: Are people left feeling like all things are new, and that life could be ever-better going forward?" They warmly reflect on my family's time there last summer, and simply say, "We just enjoy being with people in whatever part of their life they are in and giving them whatever in their mind and spirit needs an identity upgrade." I affirm, "You all certainly give a lot of love and hope to people, and really that's what's at the center of the Gospel."

John gives an "Amen" and I continue. "I'm actually going in that direction as Bee The Change has evolved over the past years. Not just to the healing of the outer landscape, but of the inner as well, and how the two intimately interact. I believe we as humans are hardwired to live in a beautiful way. Being in this place of natural beauty and abundance at your retreat center certainly helps to bring that out. What are your hopes going forward? Having more long-term volunteer/interns? More workshops and classes?" John and Audrey finish, "Probably all of that, and more Jesus gatherings, worship gatherings, permaculture workshops, and just retreats for people."

I so appreciated John and Audrey's time and loving intention. While living and working beside them in August 2020, I recognized them as another couple who have shown me true abundance. You'll remember the West family in Albion, New York. The Dutton's are nearly half their age and have, like them, embraced these two beautiful ways of seeing and finding truth in the world: Christianity and participating with & reading nature. This journey has come full circle.

Audrey Dutton launched a podcast entitled *Eat Me Drink Me: Exploring Jesus' Wonderland.* This title is in reference to the Lord 's Supper where bread and wine were eaten and drank in ceremonial remembrance of Christ. In this podcast, she explores the truths which Jesus taught and lived. She also speaks with individuals who tell of the change God has wrought in their lives, and how they have felt the Lord living and loving through them. This podcast has had ever-increasing success in the years following. John and Audrey are both remarkable individuals, always offering their creativity to beautify both the physical and spiritual world.

This journey for pollinator protection, as does life itself, continuously evolves. From land protection to land regeneration and healing. From the healing of the Earth to the healing of the infinite corridors of the human soul. From organic agriculture to permaculture. I do not see them as steps one upon another, where one may feel inclined to disregard the previous steps as "less true". I view them as integral weaves of the same fabric. As we discover new truths, we must honor the truths already attained, as they have opened our eyes to and prepared us for the new. This is just as the ascension of one mountain's peak reveals yet another to climb.

The search for knowledge and truth blesses us two-fold. I have reveled in both the journey and the destination. This bee has flown its hive, gathered goodness from myriad flowers, and returned.

17

Epilogue

Honoring the Indigenous, Living and Passed

In my travels by bicycle my foot rarely has set upon the land, yet I have experienced the land and feel an ancient impulse to honor it. I ponder, as I write, on the phrase "Old things are done away, and all things have become new". In the morning twilight of New Year's Day, I relate a profound experience I had in the Mojave Desert.

Breathing in the land, a pause is taken in the timeless space between night and day. . . A holy emptiness is felt, as though the world were waiting, having not yet taken form. In the first hour of the first day, the space seems deeply serene. Desolate. Sacred.

As the scent of sagebrush drifts along the infinite largeness of sky, I feel to call it "the smell of solitude". Traveling along the flat, imperceptibly uprising land, the shape of Joshua Trees begins to come into view, black figures upon a translucent white-orange sky. As the day arises into being, their lone silhouettes seem to take on their full, actual form. The increasing light of morning seems to transform them from shadowy concepts into full-fledged trees, right before my eyes.

147

The land was beautiful as always but seemed sad as well. I now know what this sadness was . . . and is still. It is an unmet need, a yearning even: a longing of the land for us to come back and make her our dwelling place – To dwell as have the indigenous people upon this land. They cherished these forests of Ironwood, Palo Verde, Willow, Joshua Tree, Mesquite, Pine, and arid grasslands (along with Greasewood and Ocotillo) for thousands of years. Many of their descendants, among them the Western Shoshone, Yuhaaviatam, and Gabrieleno tribes, still tend this land and dwell here.

But what is it to dwell? Perhaps a counterpoint would be helpful: Is it dwelling in a place where we hide from the sun and wind, where our food, our clothes, our shelter, even our water, comes from afar, from 'not-here'? To dwell, then, could be letting the land – with its resources immediately around us – take care of us, and reciprocating that care in our own small way.

For hundreds, in some cases thousands of years, the land patiently waits for us. Patiently, because a new era to us is but a New Year's Day to Her. She awaits our return, to make Her our dwelling place. One day, She knows, we will tire of the story of our separateness, of our ascent above Her realms. We might also feel God's compassion and tender mercies, and of His desire to care for us, and perhaps of *His* sadness as He sees His children strive to find happiness and fulfillment outside of the overwhelming goodness He has offered us, both within nature and through our own human nature.

We will then feel the ten-thousand outstretched arms and will sense the power of the old stories: of being part of a fabric, woven together, man and the world; and of living in a world surrounded by innumerable beings, even living souls. Earth knows our hearts will be won again, and we make her our dwelling place, our eternal Home.

As we return, we will come to remember something ancient and forgotten. We will remember, through our experience, that to honor her means to tend this garden of Earth, take good care thereof, and to multiply and replenish her (Genesis 1:28).

Who can imagine what we might learn about nourishing and dwelling among those of our human family as we take these "original instructions" to heart? Mankind perishes not only from lack of food and medicine, but from a dearth of community and belonging, or a famine of kindness and honor. What healing of the inner landscape might take place as the outer landscape repairs and revitalizes, having ever more to give back? As our planet returns to its prior enormous capacity of yield (relative to the ten-percent world we know today), could there finally be enough food and medicines for all, so that all are mentally as well as physically nourished? Many have dreamed of, or envisioned a world where food is free: abundant to the point both socially and ecologically where it's worth cannot translate to a monetary exchange.

This is the world that once was – Those of the Judeo-Christian tradition would have called it the *Garden of Eden*.

Acknowledgements

Many books are the result of countless gifts from myriad individuals. This book is no exception. I am humbled at the thought of how many have given to me before and during this journey, having had no thought of what they would receive in return. Their charity, in their appointed time and season, has been boundless.

First and foremost, I want to thank everyone who lent their voices and consent for me promulgate this critical information on pollinators and what we might do to contribute to their success. As a matter of respect to First Nation peoples, I'd like to recognize Cherokee ancestral land as my current environs as I finish this book. I am grateful to be able to dwell here on this land and am determined to leave it better than I found it. I thank Scott Mann of The Permaculture Podcast who made some of the interviews and visits possible as he connected me with some whom I interviewed. There are individuals such as John Coffey and Clifford Scott, whose obsession with the natural world has provided inspiration always and tutorage often.

Then there are those whose financial / material contributions greatly helped, such as Pam of Smoky Mountain Bicycles (now closed) who donated bicycle jerseys and shirts, and Paul Chew of Otto, North Carolina who made a financial donation to the cause. I also express gratitude to the Congaree, South Carolina local chapter of the Sierra Club who welcomed my authorship and increased my confidence in sharing my words with the world. As noted, one of my articles published in the *Congaree Chronicle* became one of the chapters of this book. In addition, I would like to give a shout out to Ortlieb bicycle panniers,

Slumberjack sleeping bags, KHS (road bicycle manufacturer), and Ride Bikes of Charleston, South Carolina (vendor of the KHS Vitamin B).

Though a trip by motor vehicle would have sufficed, I have long discovered that bicycling is my muse and has given me countless moments of inspiration where it seems that words and phrases – sometimes even whole paragraphs – come without effort, as if an endowment from the Universe. Keeping close to the Earth and exposed to the elements (in this case, a road bicycle by day and a sleeping bag under the stars by night) to me was not only essential to the literary experience that unfolded in this chronicle-research-type book, but complimentary to the authenticity of my earth-steward message.

My utmost thanks to Tim Havens and Scott Mann for their selfless service in editing and proofing this book. I could not have published without their guidance.

Of course, this ritual of gratitude would fall short without a mention of my father, Wayne Jerome Kotab. While we cannot raise our children with perfection, there are some aspects we can perform perfectly. What he did with utmost brilliance, assuring it would benefit my life, was teach me a deep awareness and appreciation for the natural world. While he was certain to teach about how important God, Jesus, and Holy Spirit are, he left no room for doubt that the wilderness, a product of the mind of God, was imbued with His wisdom and power. So thorough was his teaching that when I got to tenth grade in biology class, concepts such as *niche, ecosystem services, watershed,* and *biome* were already known to me and had, since a child, colored my perception of the world. My gratitude for his showing me nature as a reservoir of spiritual power runs deeper than I can tell. It, along with my Lord and my God, has redeemed me.

I am compelled to confess things as they are: The world is a place of remarkable generosity, from beings human and non-human alike.

Bibliography

Arigona, Frank. "Pathways to Intentional Communities" and "Crash on Demand". Narrator: Frank Arigona. Interviewee: David Holmgren. *Agricultural Innovations.* Web. 6 December 2021.

Barreto, Juan Delcet. "CDC's Efforts to Combat Zika in Puerto Rico Hampered by a Legacy of Mistrust". *Union of Concerned Scientists.* https://blog.ucsusa.org. Web. 12 December 2021.

Bradbury, Ray. *Fahrenheit 451.* Gallimard Education. 2000. Print.

Campbell-Mcbride, Natasha. *Gut and Psychology Syndrome (G.A.P.S.).* Medinform Publishing. 2010. Print.

Carson, Rachel. *Silent Spring.* Mariner Books, Houghton Mifflin Company. 2002 (Original 1962). Print.

Goulson, Dave. "Review: An overview of the environmental risks posed by neonicotinoid insecticides". *Journal of Applied Ecology.* Volume 50, Issue 4 (2013). British Ecological Society. Web. 6 December 2021.

Cane, James et al. "Pollination Value of Male Bees: The Specialist Bee *Peponapis pruinose (Apidae)* at Summer [and Winter] Squash (*Cucurbita pepo*)". *Environmental Entomology.* Volume 40, Issue 3 (2011). Oxford Academic. Web.

Esquivel, Parys, and Brewer. "Pollination by Non-*Apis* Bees and Potential Benefits in Self-Pollinating Crops". *Annals of the Entomological*

Society of America. Volume 114, Issue 2 (2021). Oxford Academic. Web. 7 December 2021.

Hemenway, Toby. *Gaia's Garden*. Chelsea Green Publishing. 2009. Print.

Holmgren, David. *Permaculture Principles and Pathways Beyond Sustainability*. Melliodora Publishing. 2002. Print.

Holzer, Sepp. *Desert or Paradise: Restoring Endangered Landscapes Using Water Management*. Chelsea Green Publishing. 2011. Print.

Holy Bible. Geneva Translation (GNV). 1560. Print.

Kimball, Kristin. *The Dirty Life: A Memoir of Farming, Food, and Love*. Scribner. 2011. Print.

Lu, Chensheng et al. "*In Situ* replication of honeybee colony collapse disorder". *Bulletin of Insectology*. Volume 65, Issue 1 (2012). "Sub-Lethal exposure to neonicinoids impaired honeybee's winterization before proceeding to colony collapse disorder". *Bulletin of Insectology*. Volume 67, Issue 1 (2014). Web. 6 December 2021.

Matsumoto, Takashi. "Reduction in homing flights in the honeybee *Apis mellifera* after a sublethal dose of neonicotinoid insectides". *Bulletin of Insectology*. Volume 66, Issue 1 (2013). Web. 7 December 2021.

McKinnon, J.B. "A 10 Percent World". *The Walrus*. https://thewalrus.ca/a-10-percent-world/. Web. 19 January, 2022.

Muir, John. *The Mountains of California*. The Century Company. 1894. Print. (All quotes including the two paragraphs from 'The Bee Pastures')

Partap and Ya. "The Human Pollinators of Fruit Crops in Maoxian County, Sichuan, China". *Mountain Research and Development*. Volume 32, Issue 2 (2012). BioOne Complete. Web. 6 December 2021.

Pollan, Michael. *The Botany of Desire*. Random House Publishing. 2002. Print.

Stamets. Paul et al. "Extracts of Polyphore Mushroom Mycelia Reduce Viruses in Honey Bees". *Scientific Reports*. Volume 8, Article 13936 (2018). Nature. Web. 7 December 2021

NOTE: The 2018 study was released years after the 2015 Vancouver conference referenced in "An Apple A Day". However, Stamets and his team were doing preliminary work with polyphore mushroom extracts during that time and found the initial results very promising, enough to tell of their success at the conference.

Thorpe, Frankie, et al. *California Bees and Blooms*. University of California. 2014. Print.

Thoreau, Henry David. *Life Without Principle*. Boston. Print.

Walker, Richard Allan. "It's our identity, our culture, our everything". *The Salish Current*. https://salish-current.org/2024/11/06/its-our-identity-our-culture-our-everything/ . Web. 10 November 2024.

Warren, Robert J. "Ghosts of Cultivation Past - Native American Dispersal Legacy Persists in Tree Distribution". *PLOS One.* October 31, 2023.

Multiple Authors and Articles. *Worldwide Integrated Assessment on Systemic Pesticides on Biodiversity and Ecosystems.* Volume 22, Issue 1 (2015). Web. 7 December 2021.